Delfina Cuero

Delfina and Her Friends

From left to right: Isabel Thing, Rosalie Robertson, Matilda Osuna, and Delfina

Ballena Press Anthropological Papers No. 38
Editor: Sylvia Brakke Vane

Delfina Cuero

Her Autobiography
An Account of Her Last Years
and
Her Ethnobotanic Contributions
by
Florence Connolly Shipek

Ballena Press
823 Valparaiso Avenue
Menlo Park, Ca 94025

Ballena Press

General Editors: Sylvia Brakke Vane
                     Lowell John Bean

Volume Director: Karla Young

Ballena Press Anthropological Papers Editors:

                     Thomas C. Blackburn
                     Sylvia Brakke Vane
                     Lowell John Bean

Library of Congress Cataloging in Publication Data

Cuero, Delphina, ca. 1900-1972
  Delphina Cuero: her autobiography : an account of
her last years and her ethnobotanic contributions / by
Florence Connolly Shipek.
  p. cm. -- (Ballena Press anthropological papers :
no. 38) ISBN 0-87919-123-6 : $19.50. -- ISBN
0-87919-122-8 (pbk.) : $12.50
1. Cuero, Delphina, ca. 1900-1972. 2. Diegueno
Indians--Biography. 3. Diegueno Indians--Social conditions.
4. Diegueno Indians--Ethnobotany. 5. San Diego County
(Calif.)--Biography. 6. Mexico--Biography. I. Shipek,
Florence Connolly. 1918- . II. Title. III. Series.
E99.D5C84 1911
979.4'9800497502--dc20
[B]   91-20493
                                       CIP

Printed in the United States of America
Seventh Printing

TABLE OF CONTENTS

LIST OF ILLUSTRATIONS

FOREWORD TO DELFINA CUERO

Ballena Press is very pleased to present this composite work, which completes what Delfina Cuero, Florence Shipek, and Rosalie Robertson began when they compiled Delfina's autobiography and had it published by Dawson's Book Shop in 1968. The book promptly went out of print, and was reissued by Malki Press in 1970. It has become a classic, a favorite of teachers and their students, as well as of the general public. There has been a demand for its republication ever since it went out of print a second time several years ago.

For this Ballena edition, Florence Shipek has written an account of Delfina's last years, and has organized data gathered in two ethnobotanical field trips into the format of an ethnobotany. The book as a whole therefore has a new title, *Delfina Cuero*, and contains the familiar autobiography, plus the new material. We hope it will not only please a new generation of readers, but also help bring about greater public understanding of Delfina's people and their problems.

PREFACE

Several years ago, Rupert Costo, president of the American Indian Historical Society, asked me to review this book for the *Indian Historian*. I am indebted to Rupert, himself a California Indian and a Cahuilla, for calling my attention to the most significant publication in several decades on the Southern Diegueño Indians. The story of Delfina Cuero not only provides anthropologists, historians, and others with new information on aspects of aboriginal life among these people which has been neglected in the literature, but it is also a uniquely human and moving document that eloquently captures the destruction of a culture.

My enthusiasm for this book was such that in my original review I expressed the wish that it might become available for a wider audience. As a publication of the Baja California Travels Series published by Dawson's Book Shop under the general editorship of Edwin Carpenter and Glen Dawson, the book was limited to 600 copies that were almost immediately disposed of to collectors and libraries. Malki Museum Press is grateful to Dawson's Bookshop for permission to reprint Delfina's autobiography in a paperback edition designed for a more general readership. Mrs. Florence Shipek, who compiled the book, has asked that all royalties from this reprint go to Mrs. Cuero and her Indian interpreter, Rosalie Robertson.

In reading this work, one is struck again and again by the courage and endurance of Mrs. Cuero, one of only a few members of her people who survived the decades of cultural stress upon California's coastal Indian populations. Mrs. Cuero records the tragic story of displaced peoples who were never enrolled on a reservation. She tells of Indians who lived in San Diego County until the early 1900's, surviving from hand to mouth, belonging nowhere, owning nothing, unrecognized as part of governmental responsibility and exploited as the cheapest form of labor supply (paid in food and goods only) by residents of San Diego, until finally, being no longer of use, they were forced family by family to cross the border into Baja California, where they were accepted by relatives in refugee villages. Here they managed to subsist by hunting and gathering and occasional wage labor for Mexican ranchers.

In short, the final result of the neglect of these people was to relegate them to the status of Orwellian nonpersons. Anthropological literature tells us little about them. Records of them do not exist. Consequently, although born within the territorial United States, rightful claimants to the obligations and privileges of citizenship, they are not recognized as such. It is Delfina's urgent plea that her people be allowed to come rightfully back to their homes—that she herself be permitted to come home so that she can live out the final years of her life in the comfort of the heartland of her youth.

For this to be accomplished, she has carefully demonstrated that she was in fact born and raised in San Diego County, by reconstructing the culture of her youth. The sheer human drama of this poignant autobiography will be its most attractive feature for most readers. But scholars concerned with California's social history and those engaged in ethnography and the reconstruction of aboriginal life—specialists in culture change—will find fresh and valuable data for continuing investigation and research.

For the ethnographer, the information on food collecting, hunting, and fishing, along the coastal regions of California, will prove invaluable, being specific in terms of what and how, who and when. Information on trade relations between Indian groups, ceremonial participation, leadership selection and roles is presented for the first time. Interpretations by Mrs. Cuero of oral literature, ritual obligations, and numerous other bits and pieces of data, emerging naturally in the context of the autobiography, will be subject to long analysis by those who are reconstructing aspects of Southern California Indian life.

The document is particularly valuable for the student interested in cultural change and persistence. One can trace the diminution of ritual through time. Delfina Cuero tells us who was the last person to hold a ritual and how it differed before that time, and why it is no longer practiced. The precise manner in which she recounts the interdependent relationship between ritual and food resources, so characteristic of California Indians, demonstrates clearly the close integration between environment and religious expression which has long been a source of study and conjecture for cultural ecologists.

There is so much more, providing insight into what happens to the poor and culturally distressed strata of a conquered people, that a volume might well be written on the interpretation of this work. That this book is but a part of the rich legacy of these people, now being gathered by its editor, Mrs. Shipek, is a great encouragement to students of Indian life.

Mrs. Shipek's tightly written introductory essay provides the reader with the legal background out of which this study came and the field procedures utilized in its collection. In placing the autobiography in historic and ethnographic perspective, she makes her own fresh contribution to our knowledge of the Southern Diegueños. Particularly significant has been her discovery of new lineage names and aboriginal locations of these lineages. Her brief discussion of the transition of Indian names into variations of Spanish and English is, I believe, the first to be published for the Southern California area.

Mrs. Shipek has been assisted in her work by Rosalie Pinto Robertson, her interpreter, who also helped her to collect materials, and Dr. Margaret Langdon, a linguist at the University of California at San Diego. Dr. Langdon has provided a pronunciation guide and aided in the identification of language differentiation in the area. Her guide will be of aid to many future ethnographers working among the Diegueño.

In making this reprint available, the board of trustees of Malki Museum shares the hope of Mr. Glen Dawson that it achieve the readership desired, becoming available to Indian people, students, local libraries, and schools and museums. In particular it should prove most useful as supplementary material for teachers of California Indian studies and an interesting addition to the usual texts assigned to pupils.

Lowell John Bean

California State College
Hayward, California

INTRODUCTION

IN THE course of research for the Mission Indian Claims Case, Docket 80, it seemed advisable to locate any surviving Diegueño Indians who had formerly lived along the San Diego coast. After ascertaining that there were none in the reservation villages, I explored the hint given by Kroeber[1] and Drucker[2] that surviving coastal Indians might be in Baja California. Judge Benjamin Hayes[3] had recorded in 1870 that the Indians of Old Town and San Diego Mission had recognized as Captain an Indian who lived in the mountains south of Tecate, at Neji in Baja California.

It has long been recognized that a linguistic and cultural division existed in the Diegueño or Kumǝya·y peoples between what has variously been called Eastern and Western or Northern and Southern Groups. Diegueño is the Spanish term for the Kumǝya·y people and means those who were attached to the San Diego Mission. Recent linguistic and social evidence appears to indicate that possibly a threefold division, Northern, Central, and Southern, would more accurately describe the Diegueño situation.[4] The Central group now occupies those reser-

[1] Kroeber, A. L., *Handbook of the Indians of California*, Bureau of American Ethnology, Bulletin 78 (Washington, 1925), page 725.

[2] Drucker, Phillip, *Culture Element Distributions: V Southern California.* University of California Anthropological Records, Vol. 1, no. 1 (Berkeley 1937), page 5.

[3] Wolcott, Marjorie Tisdale, *Pioneer Notes from the Diaries of Judge Benjamin Hayes, 1849-1875* (Los Angeles, 1929), pages 296-298.

[4] Private communication from Dr. Margaret Langdon, Linguistics Department, University of California, San Diego.

vations scattered across the southeastern third of San
Diego County. The Southern Diegueño occupied the territory from Torrey Pines south along the coast and extending at an angle inland and southerly to Jamul and
Tecate and beyond, approximately fifty miles south of
the border into Baja California. Thus, general relation-
ships for the San Diego Bay and Mission Valley Indians
were closer to the south and southeast than they were
to the north or directly east. Within the overall Diegueño
area and the three general linguistic divisions, each village had its own variant dialect; and closely adjoining villagers could understand each other more readily than
could those from more distantly located villages.

My interpreter, Mrs. Rosalie Pinto Robertson of Campo Reservation, went into northern Baja California and
renewed her friendships with Indian people whom she
had known as a child, including some who were related
to her and some who had relatives at Jamul village. Slowly
but surely we located one Indian after another who had
formerly lived around San Diego—Indians who had been
born either in Mission Valley or immediately to the east
of the Valley and whose parents had come from Mission
Valley. The Diegueño Indians whose home territory had
always been northern Baja California and who were on
their own land were quite emphatic when identifying
those who had come down from the north.

In replies concerning an individual's origin, Indians
are quite literal and specific in naming the actual village,
camping spot, or place of birth. I asked a very old woman
who had always lived in the Jamul-Barrett-Dulzura area
if she knew Delfina Cuero, the subject of this autobiography. This woman replied that she had been about 10

years old when Delfina was born and that her mother had been a part of the same group as Delfina's parents. When I asked where Delfina was born, instead of telling me, she had me drive to Jamacha, where she pointed to an old grove of trees and said, "In a little Indian house under those old trees."

It was discovered that until 1900 and 1910 many Diegueño Indians had lived in Mission Valley and in various places around San Diego. A favorite spot was between 13th and 17th around K Street. Other Indian living areas were: on the bay at the foot of Fifth Street, along the Silver Strand, at the foot of Rose Canyon, along Ocean Beach, around the edge of Mission Bay (False Bay), and all up and down Mission Valley. Each of these locations has been corroborated independently by non-Indian "old timers" in San Diego.

During this same period of time, in addition to the Indians on the reservations of San Diego County, there were non-reservation Indians camped throughout Lakeside, El Cajon, Monte Vista, Jamacha, Otay, and in all the little mountain valleys of the San Diego back country. The older ranchers of the region have confirmed the stories of the Indians concerning the residence locations of the Indians.

By 1910 the white and oriental populations of San Diego were increasing and filling Mission Valley with small farms and the city area with houses and business. The Indians gradually moved out of the coastal regions. They were technically "squatting" and did not own the land they occupied. As they had to move seasonally in search of food, they were not even permanently in one squatting location on a year-round basis. By 1920 most of the Indians had left the San Diego-Mission Valley area.

During this time and including the 1920-1930 period, the back country non-Indian population was growing also. Some large holdings were divided and the public lands were filled. Again, the squatting Indian was told to leave; the small land holder had no place for Indian families on land he had purchased. It was during this time, according to the Indians, that John Spreckels owned Jamul Rancho and gave a small corner of the Rancho to his faithful Indian ranch hands as a home where they could safely remain. The land title records show that the Coronado Beach Company, in which Spreckels had a controlling interest, deeded this corner of the Rancho to the Roman Catholic Bishop in 1912 for the purpose of maintaining an Indian graveyard and the approaches thereto.

The majority of Indians kept moving, looking for food, working for a rancher a while here and a while there, and then looking again when Indian labor was no longer needed. Except for a few who had married into reservation families, the majority of these San Diego Indians did not feel free to move into the Indian villages on the Southern California reservations. They were not related, nor invited. They also spoke a slightly different dialect. These Indian people gradually moved from the San Diego area of high non-Indian population into an adjoining region of low non-Indian population, northern Baja California. Closely related dialects were spoken in the two areas and the Indians would also be able to continue their traditional type of life and economy. Baja California was the natural refuge.

The Indians had no formal education, knew nothing of non-Indian laws or that they had any rights whatsoever. They knew merely that in the past they had moved freely throughout Kuməya·y territory attending funerals, ceremonies, and going to major food-gathering grounds in the proper seasons. They knew nothing of an international border which cut their territory through the middle. They knew only that Indian existence was still continuing in the southern portion of their territory. During interviews with Indians from both sides of the border, it became apparent that only in the last twenty years have the majority of unschooled, non-reservation Indians become conscious of the border and felt restriction on their freedom of movement within Diegueño territory as they visited relatives, attended ceremonies, gathered acorns and pine nuts, and even changed their residence in either direction.

In contacting the elderly Indians, my purpose was to obtain accurate information about Indian life in the San Diego coastal section. Those Indians who were identified as from San Diego were questioned intensively concerning their life: where they had lived, how, what plants had been gathered and where, what fish and shellfish used, what social and religious activity had occurred, what was their total activity in the coastal region. Permission to bring some individuals back across the border for a few days was obtained. We visited various localities which they had described to me, gathered plant specimens, and checked fish and shellfish identifications; all remarked on the changed landscape (due to cut and fill operations) and the strange new trees. Each informant was constantly tested and cross-checked for accuracy and reliability. For example, each Indian on being taken to the Campo Santo (cemetery) in Old Town, San Diego, remarked that it was much smaller than it had been. Local records indicate that it was reduced in size when the road was paved. Another example is that lineage names for Indian families which had formerly existed in this area were obtained. Some of these names had never been recorded anthropologically prior to this research. Subsequent examination of the very earliest San Diego Mission baptismal and marriage registers revealed that amongst the earliest converts to the Mission were Indians with these, now extinct, but remembered, names. The informants' only source for learning this information would have been participation in the now extinct San Diego Indian community. These people could neither read nor write; some understood a little English but none spoke it. Some had a minimal amount of "ranch" or "kitchen" Spanish. The services of Mrs. Rosalie P. Robertson as interpreter between Kumǝya·y and English were essential at all times.

It must be kept in mind that unwritten custom in a small, tightly knit society normally changes but slowly and is always a much stronger force than is written law in a complex urban civilization. The San Diego Indian community had been forced by circumstances beyond its control to make radical changes in custom. In the course of this research, the survivors were asked to break one more custom, one which was difficult and mentally painful to break: a taboo on the discussion of, and the naming of, deceased relatives and friends.

There are not many living Indians, even counting children and grandchildren, who fit into the category of unwilling refugees from San Diego County. In addition to Delfina, those left of the many that went south are:

Albert Ames of San José del Zorro; Matilda Osuna, whose children and grandchildren are still in Tecate; Olojio Thing of San José near Tecate; Eleanor Quihas and Fernando Quaha of Ha-a; and their descendants. The Lamop family at San José de la Zorro was from the San Diego Mission.

Much of the ethnological data presented in this auto-biography can be found in Kroeber[5], Gifford[6], and Spier[7]. Many details which Delfina gives add to previously re-corded data. Some are variations; but as Delfina said, "This is what happened to me; what I was told. They may have done it differently in other places and at other times. Many things are different today from when I was young. I tell only what I know."

Indian lineage, or clan, names are presently used in several ways. The most direct is the lineage name written in Spanish orthography with the original Indian sounds shifted to the closest Spanish sounds. Another form of usage is to translate the Indian name into the Spanish or English equivalent in meaning. Some names have been changed into a Spanish name closely related in sound but not in meaning. Other Indian families have simply taken an unrelated Spanish name for convenience in usage. Closely related branches of the same family are sometimes found using different forms at the present time. Some examples are:

[5] *Op. cit.*

[6] Gifford, E. W. *Clans and Moieties in Southern California,* University of California Publications in American Archaeology and Ethnology, Vol. 14, No. 2 (Berkeley, 1918).

[7] Spier, Leslie, *Southern Diegueño Customs,* University of California Publications in American Archaeology and Ethnology, Vol. 20, No. 16 (Berkeley, 1923).

Phonetic	Spanish Orthography	Translation Spanish	English	Similar-sounding or other name.
kʷał	Quash	Cuero	Hyde	
kunʸi·ły	{ Quinnich Conhitch	Prieta	Black	
ʔəsu·n	Osun			Osuna
kʷi·xa·s	Quihas			{ Calles Rosales
miškʷi·š	Mishquish			{ Mata Mesa
kʷaxa·	Quaha			Lopez
xatʔam	Xotom			Tampo

When marriage is mentioned, Indian marriage custom as described by Delfina is meant. Indian divorce was very easy: simply separate and return to one's parents. Indian marriage and divorce were recognized as legal by the United States Bureau of Indian Affairs for California Indians until 1954, when legal responsibility for the Indian person as distinct from his land was handed over to the State of California. At that time the California Indian became subject to the same laws, duties, rights, privileges, schools, and welfare on the same basis as any other citizen of California.

Diegueño custom retains for the married woman what American society terms her maiden name. Each child acquires its father's lineage name and retains it throughout life. In only a very few cases in Southern Diegueño territory have extremely elderly women come to be known by their husbands' names after many years of marriage. Margaret Brown and Isabel Thing, both mentioned in this autobiography, are among the very few.

This autobiography is typical of the life stories of most of the Indians who had no place to call their own. The men, and those women who were not widowed with small children did not have quite as much difficulty as Delfina. All relate the same search for a place to live and the same terrible struggle to feed and clothe the children, whom they loved dearly. In spite of this desperate struggle for life, these Indians are a cheerful people. They are realistic and live each day as it comes. A free translation of part of their ceremonial mourning song reflects their attitude, ". . . things may be going well for you one day, then something happens and you are destroyed. This is the way life is. Remember, it can happen to you too!" They hold no bitterness about the past and are only looking for the chance to work and to earn enough to feed and clothe their children. They feel that they have survived through the crisis and that there will be a future for their children as the children are learning the new ways.

In Delfina's life is seen the destruction of Indian self-sufficiency on the land, of Indian society, culture, and religion. Visible also is a much, much slower pace of Indian integration into modern society.

Delfina was willing to cooperate completely in my research because she hoped that if I knew the details of her life I might be able to find some document which would prove that she came from the San Diego area, enabling her to return to the United States permanently with her children and grandchildren. She wanted a search for any written record that might have been made by someone for whom her father worked. She also knew that some old Indians had baptismal records and wondered if such a record might have been made for her when she was born. Unfortunately no church records exist for the Diegueño Indians between 1900 and 1919, the time during which she might have been baptised. The Catholic Church in El Cajon, which had charge of the Indians from 1900 on, was destroyed by fire in 1917 and the records burned. Practically no other records were made of or for Indians except the Church records or the Bureau of Indian Affairs records for the reservation Indians.

After my search for written records proved fruitless, Delfina then expressed the hope that her full cooperation with the ethnological research for information about the Indian usage and history in and about San Diego might in itself prove the truth of her claim to a San Diego origin. She hopes that this testimony can replace the normal documentary proof of United States origin.

Delfina Cuero came to Jamul for her father's funeral and also for the funerals of several other relatives. She has also come over to visit and help care for sick friends. She has been interviewed each time. Delfina has been an excellent informant with a remarkable memory of places, plants, and activities. The only areas in which she had any difficulty locating herself were those sections which had been cut, filled, and covered with buildings, or were otherwise altered from the original natural contours of the land. As remarked about the informants in general,

Delfina neither reads nor writes any language; she speaks no English and only what might be called household or ranch Spanish. There is no way in which she could have acquired the detailed explicit knowledge of the San Diego area except from her own personal experience.

Delfina is one of the strongest persons whom this writer has had occasion to interview and to know. She is always laughing, joking, cheerful, and seeing the humorous side of life. Yes, she has had a hard time, been poor and hungry many times, but life is still basically good. She laughed many times as she recalled her childhood and she smiled about the pleasant parts. Living with hunger had not diminished her enjoyment of her childhood, her family life, and her playmates. Her husband, the father of her children, had been a good man and she had been happy with him. The worst is now past; she and her children have survived. She is looking hopefully into the future and praying for better things for her children and grandchildren.

Mrs. Rosalie Pinto Robertson, who translated from Kumaya·y into English, and I have attempted to make the words and ideas adhere as closely as possible to Delfina's original expressions. Throughout the autobiography, my explanatory notes are placed in brackets.

I wish to express my appreciation to all who have participated in this autobiography:

To Delfina for her willingness to cooperate in the search for the old Kumaya·y life.

To Mrs. Rosalie Pinto Robertson, interpreter and friend, who has patiently educated me in Diegueño society and culture.

To Dr. Margaret Langdon, linguist, University of California, San Diego, for the correct spelling of Kumǝya·y words and for reading this manuscript.

To Mrs. Elisabeth C. Norland, San Diego Museum of Natural History, for her identification of plant specimens.

FLORENCE C. SHIPEK

San Diego, California

PRONUNCIATION GUIDE

In 1968, when the first edition of this book appeared, there was only one systematic writing system available for Diegueño languages, that which was to appear in Langdon (1970). Since then, as a result of the interest of several Indian communities in San Diego County in writing their languages, a practical orthography was developed and first described in Couro and Langdon (1975). The native words used in the body of the book, therefore were written in the first technical orthography and remain in that style in the new edition. However, since the practical orthography has been widely used recently, not only in linguistic publications, but also in newspapers and other places, the native terms used in the Epilogue conform to this newer practice. The pronunciation guide below incorporates both practices and should be interpreted as follows: symbols under the heading **I** are the technical orthography of Langdon (1970) and the symbols under **II** represent the equivalent of those under **I** in the practical orthography of Couro and Langdon (1975).

Although there is, strictly speaking, no exact English equivalent for any of the Diegueño sounds, a reasonable approximation of native words can be achieved by using the equivalencies listed below for each of the symbols. Where no English equivalent exists, examples are taken from the best known language in which the sound does occur and additional phonetic instructions are given. The symbol "·" following a vowel, or its equivalent in the practical orthography--the doubling of the vowel symbol--indicates that the vowel is long. Since more than length is involved, however, the key below lists long and short vowels separately, with English equivalents for each. Finally, and this is not marked in the transcription, each word is automatically accented on the last syllable.

Langdon, Margaret. A Grammar of Diegueño: the Mesa Grande dialect. University of California Publications in Linguistics 66 (1970).

Couro, Ted and Margaret Langdon. Let's Talk 'Iipay Aa. An Introduction to the Mesa Grande Diegueño Language. Malki Museum Press: Banning, CA., and Ballena Press: Ramona, CA (1975).

I	II		
a	a	=	*u* in 'but', or *a* in 'sofa'; when followed by 'y' as in ?u·tay, 'Otay', it is more like *e* in 'met'
a·	aa	=	*a* in 'father'
c	ch	=	*ch* in 'church'
	e	=	*e* in 'father'
i	i	=	*i* in 'pit'
i·	ii	=	between the *ee* of 'keep' and the *a* of 'mate' (closer yet to French *e* as in *ete* 'summer')

k	k	=	*k* in 'keep'
k^w	kw	=	*qu* in 'quite'
l	l	=	*l* in 'loom'
ł	ll	=	the same sound a 'l', but without vibration of the vocal cords. It is the Welsh sound spelled *ll* as in 'Lloyd'
l^y	ly	=	*li* in 'alien'
ły	lly	=	the same sound as 'l^y', but without vibration of the vocal cords. In other words, 'l^y' is to 'ł^y' as l' is to 'ł'
m	m	=	*m* in 'mother'
n	n	=	close to *n* in 'no', but with tip of tongue touching lower edge of upper teeth (closest to French *n* in *nid* 'nest')
n^y	ny	=	*ni* in 'onion'
p	p	=	*p* in 'Peter'
r	r	=	Spanish *r* in pero 'but' (i.e., trilled)
s	s	=	*s* in 'Sam'
š	sh	=	*sh* in 'ship'
t	t	=	French *t* in *tu* 'you' (tip of tongue touches upper teeth)
ṭ	tt	=	*t* in 'too'
u	u	=	*u* in 'put'
u·	uu	=	between *oo* in 'boot' and *oa* in 'boat'
w	w	=	*w* in 'we'
x	x	=	German *ch* in *acht* 'eight', or Spanish *J* in *Juan*, 'John'
x^w	xw	=	*x* followed by *w*
y	y	=	*y* in 'yes'
ʔ	'	=	a catch in the throat, often occurs in English exclamations, i.e., at the beginning of 'ouch!' or between the two o's of 'oh-oh!' when indicating surprise or reproof

MARGARET LANGDON
University of California
at San Diego

THE AUTOBIOGRAPHY

THE AUTOBIOGRAPHY

OF

Delfina Cuero

✣

MY NAME is Delfina Cuero. I was born in xamca·
(Jamacha) about sixty-five years ago [about
1900]. My father's name was Vincente Cuero, it
means Charlie. My mother's name was Cidilda Quaha.
They were Kumɘya·y Indians. My father and my mother
came from mat kuṭap (Mission Valley in San Diego).
Now they are dead and buried in xamu·ɬ (Jamul) Indian
Cemetery, both of them.

My father's mother was Martina Kunʸi·ɬʸ (Quinnich).
His father was kari·ti· kʷaɬ (Karretay Cuero). My mother's father was xamlu·l kʷaxa· (Kamlul Quaha) and her
mother was xɘsu·s miškʷi·š (Jesus Mishkwish). They
were all born in Mission Valley. They were not raised in
the Mission San Diego. There was nobody there any
more; just some Indians and their families living there
and other Indians all up and down the Valley. My
mother's mother is buried somewhere in Mission Valley,
but I don't know where.

My father and mother left Mission Valley, they told
me, when a lot of Chinese and Americans came into the
Valley and told them that they had to leave. They did
not own the land that their families and ancestors had
always lived upon. They moved east into ʔɘwi· ka·kap

23

(Mission Gorge). There were many Kumɘyaˑy villages all through that way, up the canyon, and in Lakeside and El Cajon valleys. There was a big xaˑtuˑpiṭ [Padre Dam, built by the Mission Fathers] that they talked about where there was a village also. My mother and father went to one village, and then the next one, and on and on. After they had to leave their own place, they lived around wherever there was work or wild food to be gathered. They lived in El Cajon, then Jamacha. That's where I was born. Later we lived in xamuˑɬ (Jamul) and then sɘmuˑxuˑ (Cottonwood or Barrett). We just worked at a place and then my father would tell me we had to move again.

The Indians had names for every little spot. Many names I have forgotten, but each name meant something about that place. ʔuˑtay (Otay) means a kind of weed that grows there, that is, a lot of that weed grows in that place. xamcaˑ (Jamacha) is the name of a wild gourd and a lot of them grow in that place. xamuˑɬ (Jamul) was named for a kind of weed that grows there. The xa part of the name means that it grows where there is lots of water. Point Loma was called mat kunʸiɬʸ (black earth) because that is how it looks from the distance.

The Indians didn't speak English in those days. The men they worked for never told them their names. But there was one old timer, a pretty old man who lived in Barrett that we worked for longer than most. I remember his name was Maxfield. He might still remember us if he is living, but I doubt it. He was pretty old then. I was very young but the whole family was working for him, my father, mother, grandfather, grandmother, and I. All doing some kind of work. I can remember, we cleared a lot of land. The men cut trees and they did many other things while we carried, piled and burned the brush. Other times while the men worked, my mother, grandmother and I used to go all over looking for wild cherries at the right time of year, or wild wheat, or different kinds of Indian food. That's what we used to eat. We lived on those things we gathered when I was little. This old man gave us some food too. The ranchers that my people worked for gave us some food or sometimes some old clothes for the work. They never gave the Indians money. We didn't know what money was in those days.

The Indian women never worked for the ranchers' wives where my family was. I heard they did it other places. We worked with the men sometimes. Most of the time we hunted for wild food, greens, honey, and things.

Maxfield gave us a little place where we could stay while we worked for him, a place to build our səmay ʔəwaꞏ [little Indian house of willows and other brush]. The men put up two posts and tied a beam between them with fibers stripped from yucca leaves. The reeds or brush were tied to the beam. It looked kind of like a small [pup] tent. We used tamuꞏ (reeds) when we could get it otherwise we used xatamuꞏ [Haplopappus sp., brush]. Then the men put four posts to make a square and on them we made a ramada beside the house.

I remember my father used to work all around El Cajon and Jamul and many places. He did ranch work. We just camped as close as we could most times. We never lived in a house. We just lived out away from the ranch houses in the brush of some small canyon. My father would pick a place where the wind couldn't hit us. In those days there was water in places that are dry now.

The Indians had to move around from place to place to hunt and gather enough food, so we knew lots of places to camp. Later on white people kept moving into more and more of the places and we couldn't camp around those places any more. We went farther and farther from San Diego looking for places where nobody chased us away.

My grandparents crossed the line first. In those days we didn't know it was a line, only that nobody chased them away from xaʔaꞏ (Ha-a, the willows) [a small Diegueño Indian village in the mountains about twenty

1   ither
                                                         own
                                                         rom-

                                                         vhat
                                                         ould
g___ __ _ p____ like that. In those days when you were with one group, you stick with that group. You can't go in with another. Most of the people we knew went down there [Baja California] hunting for different food and they found a place where no one told them to move on, so they just stayed there. Some Indians were already at xaʔaꞏ who spoke the same kumэyaꞏy language, just as we did.

When the Indians were told to leave a place, they generally just headed farther into the mountains. Pretty soon they would tell me we had to move again. We would pack up everything we had, a few saꞏkay (ollas) and baskets of dried food if we had been lucky and found enough to dry for winter, a few small stone tools, bow and arrows, xampuꞏ [throwing stick for hunting rabbits], and what little clothing we had. We would pack everything on the back of a small burro that my father had and go. My father was a good man. He was never a drinker then. He worked hard and so did my mother.

I can remember when I was pretty young, we used to go and look for xaˑlʸak (abalone). They used to show me how to find them and get them off the rocks. We used to hunt for fish, shellfish, and other stuff in the ocean and along the edge of the ocean around Ocean Beach. There are so many houses here now I can't find my way any more. [The shore line at Ocean Beach has been filled in part.] Everything looks so bad now; the hills are cut up even. I can remember coming to mat kunʸiɬʸ (Point Loma). There were not many houses then. I was little and don't remember all, but there was a lot of food here then. We had to hunt for plant food all day to find enough to eat. We ate a lot of cactus plants [Dudleya sp. and Opuntia sp.]. We ate a lot of shellfish. There were lots of rabbits there too. My grandfather used to tell the boys to eat rabbit eyes especially, because it would make them good hunters. We hunted those things until we couldn't hunt on Point Loma any more.

I remember we walked a long way to get ʔəsnʸaˑw (acorns). I know we went into the mountains. I think ʔəkʷiˑ ʔəmak (Cuyamaca) and Laguna were the places. There were lots of big acorn trees there. We would gather a lot and pack them on our backs down to the coast again.

We used to gather pine nuts right near the ocean near San Diego beyond mat kulaˑxuˑy (La Jolla or "land of holes"). If there weren't so many houses maybe I could find my way to all the places again. It wasn't far from Mission Valley to the place for pine nuts [probably

Torrey Pines, now a State Park]. The men got fish and other things from the ocean when we got pine nuts. There were a lot of səmay a·suw [vegetables or eating greens] all over near the ocean.

There also was another place where we used to go to get pine nuts, a long way from San Diego and near where I live now, on the other side, near La Rumerosa, Baja California. Those were the only two places where we got pine nuts. We used to meet Indians from all over Kuməya·y territory when we gathered pine nuts, even some Indians from the reservations.

There were always lots of other Indians living around the same way we were. The Indians ate everything possible in those days. They were hungry if they didn't.

We used to hunt for all kinds of shellfish. We would boil some to eat then. We would clean, wash, and cut the rest and spread them on the rocks to dry in the sun. If we had si· [salt], we would put that on and the meat would keep better. We used to get salt down at the southern edge of San Diego Bay.

We used a sharp rock that fitted in your hand with a point at one end to get abalone off the rocks. We used to search around the rocks at low tide for them. You have to pound the meat of abalone soft with a rock right away. Then you cut it up and set the meat to dry in the sun, just like the other fish. The xa·lʸak [abalone shell] made good dishes; other smaller shells were used for spoons.

My grandparents used to eat a small white part [the stomach] of the ci·kʷəɬʸap (starfish), but I never ate them myself. Crabs are good meat too. We ate many things that look ugly but that are good meat. I remember we caught a xilkəta·t (octopus) a long time ago; it was real ugly, but food. If you dig a hole in the muddy sand with a digging stick and put water in it with an olla, the ki·xuɫ (clams) will come up. We ate scallops too. Anything we could take, we ate. We ate miɫʸkama·w (lobster) and wee little stuff that looked like spiders, real small—ki·taš (shrimp).

There was a story about the olivella shell; they were babies that fell from the stars. They used to say: When the dipper in the sky [the Big Dipper] gets too full, it is dumped out. Then these small shells fall all around near the ocean. There was more to it but I am not a story teller and that is all I can remember. They used to name all the stars and tell stories about them, and explain why the Dipper is lying differently in summer and winter.

We caught fish and cleaned them. We took the fins, tail, and head off and used those parts to make a good soup. The eyes, especially, were good for you. I thought they were so ugly that I never cared for them but I used them to make good soup. I would clean the fish and boil it to eat. When we got a lot, we would cut it all up and dry the meat in the sun for later. We used cactus thorns on a long stick to spear fish. We also made traps out of agave fiber. We put the traps in the ocean, put a piece of rabbit meat in it, and could come back later to get the fish. We made nets out of tall grasses; ropes and nets were made of agave too. We had other ways to catch fish too, but I don't remember them all. The women made small nets and the men made big ones. I can remember the old timers talk about making kəyuš (boat) out of tamu· (reeds). They would weave them so tightly that the water could not get through.

I remember there was a little fish that came in swarms

on the beaches during high tides on some spring and summer nights [grunion or *Leurethes tenuis*]. We used to get lots of them and dry them for the winter.

They used to eat a lot of other things too, but I can't name them all now. We used to eat šamuk [an edible sea weed], a plant that grew in the ocean. We used to eat some of the birds along the shore too. There used to be a story about how the little sand fleas took care of the ocean, but I haven't heard it since I was real small.

We always carried some grinding stones and some other tools with us. A lot of stones you could pick up any place and make what you needed.

We would fix meat that same way we did fish. When there was no salt, we would split it thin and dry it in the sun. Any kind of meat we could get we used—rabbit, deer, opossum, raccoon, wood rat, anything. Opossums are good food; the meat is real good. I still eat them when I have a chance. But now I'm getting old and I care more for cottontail rabbits; they're easier to find. I still live on my Indian food when I can get it.

We used to grind pine nuts into pinole [flour mixed with water and honey] to eat. We ground most nuts and seeds into flour—wild cherry, lilac, flowers of all sorts and grass. We winnowed it and sifted it in a samilʸ (round, flat basket) after we ground it on a flat grinding stone.

ʔəsnʸa·w (acorns) had to be pounded in a ʔəxmu· (mortar), sifted, and pounded again. After it is all fine, like flour, the bitter is leached out with water. My grandfather told me that a long time ago when they didn't have anything to leach the wətuṭ (ground acorn meal), they would fix a place in the sand and pack all the acorn

meal in there real thick and then put more on top; then
soak it and soak it. Then they would scrape off the top
meal to cook it to eat. That was when they didn't have
nʸipuˑn [special leaching baskets] to use with them.

We used to carry loads on our backs with bags made of
agave fiber. We used big gourds for dishes and for storing
stuff as well as mat saˑkʷinʸ (clay dishes) and ollas, and
different shaped baskets. We made our houses out of
tamuˑ (reeds). We tied them with yucca fiber so close
and tight they held the rain out. We wove mats to sleep
on from tamuˑ and sometimes they would use the mats
as rafts.

We gathered green vegetables and roots. My mother
and grandmother taught me what to gather. We would
boil them and spread them on a rock, or a special basket,
or on a bush to dry. That would keep it clean while drying. We dried pumpkins too. I still do when I can get
them; I love pumpkins. In my garden right now, I have
to cut my own fence posts and set them and fix the
fence and water and weed by myself. I used to make
some xəmšuˑṭ (drying baskets) by twining them, but not
the good coiled baskets. I still store some greens and
other things when I can find them. The seeds of the
milkasup [wild sweet pea with red flowers] are good
food, I fix them like beans.

What I used to like a lot, my parents used to show me
how to crack and grind xaˑsił (manzanita berries) and
make a drink. With a little honey in it, it's like Kool-
aid. Before there were a lot of houses there, we could get
manzanita in Pacific Beach between Mission Bay and the
ocean.

The women had to do their work while the men

worked too. Either we do this or we starve. We had to learn how to use all these plants, what to hunt for and when. We used to go out a long way sometimes, gather the food, put it in our bags and pack it back. Then we had to clean it, dry it, and store it for winter. We always stored as much as we could find for winter when not much would be growing.

We did that when I was young and my grandmother told me they always did that—store as much as they could get for winter. The men and the women worked hard to get enough food. The greens and roots were cooked and then spread on the rocks to dry. If no rocks were near, the women would quickly twine some loose open baskets out of a tough green grass that grows in wet places. Then they would spread the cooked vegetable in them to dry.

In April and May we used to hunt over toward the desert for mescal. We would dig a pit and roast the stalk. Some Indians still do that, it is real good food.

When I was young, we had to move too much to plant anything. Always being told to leave, it was no use to plant. My grandmother used to tell me that when the Indians could live in the same place and could come back to the same place from gathering acorns and things, they would always clear a little place near their house. In it they planted some of the greens and seeds and roots that they liked, just the things that grow wild. That way, they had some food close to their house. Sometimes they put a piece of cactus near their house and it would grow. But when I was young it was no use to plant like that when we couldn't stay to get it.

We used to eat rats, mice, lizards, and some snakes,

but I don't remember what kinds. We roasted them. The little things were pounded on a rock, bones and all and then stewed. We had to eat whatever we could find to have enough food. We were hungry many times. We were poor. The old people used to eat roasted grasshoppers. They pulled the legs off and ate them. The little white worms of the bees were good food when roasted. They were real greasy and tasted sort of like peanuts; they were real good. The bees come from a long way. There is a story about how a man brought them across the water in a ship and then they got loose here. [The honeybee is a European import.]

In the old days, the people down near San Diego used to take lots of salt from the bay and trade it for mesquite beans and other things from the desert. They used to go a long way to trade for what they needed. There were no roads then, just trails, and we walked and carried everything on our backs. Dried sea food, pumpkins, and dried greens were traded for gourds, acorns, agave, and honey.

There were lots of good plants to use for soap. There was yucca in the mountains and the desert. The roots were ground and used to wash with. Down around the beach areas there is səni's [Mesembryanthemum crystalinum L.]. We used to grind the whole plant except the fruit for soap. It was very good for washing hair. We used the juice from the ripe red fruit for painting our faces. Another good soap plant that grows in a lot of places is ti·šiɬ [Sueda Torreyana Wats.]

There were a lot of "vegetables" in Mission Bay. More mud and weeds were in the bay then, like marshland. Between Old Town and Point Loma was a lot of black mud where it is dry land now. We would gather the greens and roots and boil and dry them. Seeds and things like that were stored for grinding later. The food that we gathered from the mud and around the edge of the mud was real good because it was salty. It tasted good and it kept real good.

Many Indians were camping around like that. We were not the only ones that had to get out as the other people came in. Generally a lot of relatives lived close by, just camping here and there as groups together. We were all hungry many times because we just couldn't find enough food.

What can we do, we were just here and there. We don't have nothing. Each time we moved, we had to. My father and mother would say, "Help pack up, we have to move." If we had been lucky, we had the dried food we had stored for winter. We would carry that with us when we had to move.

When I was little, I had playmates—cousins and friends—around. We played a lot. Most of my friends are dead now. There was a lot of the Calles family, Mary, Juan, and others. They lived around where we did and at Jamul. Matilda Osun was a good friend. Some I haven't seen for a long time. Matilda and I used to make dolls of rags and stuff them with manure and use sticks for legs. We made clay dolls and animals also. We didn't put them in a fire to make them hard, so they would fall to pieces after a while. We made a lot of mud dolls with stones pushed into a slit made with a stick for eyes. We made a lot of mud dolls.

We used to have wars against the boys. [laughing] I can remember how crazy we were. We used to have wars with them with xamca· (wild gourds), throwing them at each others' heads. We could have cut someone or put out each others' eyes. We didn't think about that then, we had so much fun. I've been hit many times in the head but we all had a good time. Sometimes boys and girls would be on both teams and other times it would be boys against girls. Sometimes it would end in a big fight and everybody would get mad. If someone got hurt we would get mad and have a real fight, but we were all right the next day—friends again. The gourds were round and hard, like hard balls. I can remember I used to get a licking all the time because I was always throwing things.

I was always trying to fight too. I used to wrestle with everyone. I could wrestle pretty good when I was young. I used to throw Juan Calles and all the boys and girls when we were little. The girls just had one dress and we'd fight. I'd throw them and the dress would tear and they'd cry and cry. It was the only dress they had. I tore my dress too, but I sure liked to wrestle.

We had a lot of fun with all kinds of games. We played hockey with a wa·ta·š (stick) and a ball and a goal. Our ball was made out of sticks. We shot bows and arrows and threw xampu· (rabbit sticks) at targets. My father made me a rabbit stick [a curved flattened stick similar to a boomerang] but I wasn't very good with it. The old folks made little bows and arrows for us. Even the girls used them and learned to throw rabbit sticks. Some girls were good.

We used to make sticks for horses and ride those. We didn't have any toys to play with. We made our own things. We made clay horses sometimes to play with. Even now, as old as I am, some days I'll be going and I'll see a puddle and bend down and make a face or animal's head in the mud. Just model in the clay. When I come back next time, if the rains came, I don't see it any more.

I never made baskets or ollas. I never cared to do that. My grandmother could make beautiful ollas and things, also my mother, but I think I'm lazy. When I was young I was different, I always believed in looking for plants, food and herbs, and different things. I never took time for ollas and baskets, I've always worked like a man. I've had to cut wood. I still do at my age. I guess I'm just too lazy to sit in one place. I couldn't do it. I cut fence posts for my goats last week.

Besides wrestling when we were young, we used to have fun sliding down smooth rocks. I ruined my rear end one time I slid down the rocks so much. We used to have races too. I used to outrun Fernando Quaha. He is older than I am, too. I don't know about now though, he might outrun me now.

We used to get on the highest rocks and jump down to see who was brave. We used to play rough. Sometimes some one would push another off and someone would break an arm or get hurt. But most of the time, when we played we had fun. We didn't have a lot of time to play. Most of the time I went with my mother and gathered greens to eat.

Grandfather used to tell me that in the olden days they never wore clothes like we do now: only a təpaꞏraꞏw (loin cloth) for men and bark aprons for the women and hides for winter when it was cold. They were very poor then. Grandfather used to talk about big dances they used to have when he was young. I never paid much attention. When I was growing up, Grandfather lead the kəruk dances [the mourning ceremony for all the dead]. He was the one that danced with the məskʷaˑ [image made of reeds and brush which represented the deceased in the ceremony] in the taˑkay [mourning ceremony one year after the death of a person]. Grandfather knew all the songs that went with the death ceremonies and the image. He was an important man because he knew all these things.

In the old days, the image dance was done with a bunch of deer hooves instead of a gourd rattle like they use now. They cut the ends of the deer hooves off, made a little hole in them and tied them in a round bunch with hide or a mescal string. The singer held them in his hand and shook them in time to the song, first up and down and then sideways.

The Indian men and boys used to make flutes to play in the evening. One flute was made out of a hollow stem with just the holes at the end. The man would blow at one end and put his hand back and forth in front of the other end to make music. Then they made another kind out of a hollow stem from a plant that had yellow flowers. The flute was about a foot long and they put six holes along one side to make the different sounds. I used to think it was real pretty. They stopped doing it a long time ago when I was still a little girl.

I remember my grandfather used to tell the boys if they wanted to grow up and hunt, they had to be clean. They used some kind of sage and washed and bathed in it. He used to preach about a lot of things. Old men always preached to boys, "Be Clean!" They told the boys how to live. One most important thing, a young man should stay away from a woman menstruating or pregnant. It would ruin a hunter's eyesight and he couldn't find rabbits if he didn't. It was real bad for the bows and arrows and also for the gun for such a woman to touch the weapons. It would ruin the sights of the gun and the bow wouldn't shoot straight either.

It would ruin the ladies too; it might cause them not to menstruate regularly. One week you have to stay away from your husband; also one month after you have a baby. Also you have to stay away from greasy food and eat only atole [ground wheat or corn mixed with a little water]. Stay away from salt too! Ladies had a kuca·ma·l nʸə wa· [fast house—a little hut off to one side] to live in when they needed to be by themselves.

Girls were taught these things and how to be clean by their grandmothers. If a young woman was going to make an olla, she must go off by herself. If a person with aches and pains came around while she was making the olla and just looked at it, it would not stick together good. It would just break.

They used to have a lot of rules for young children about how to behave. If strangers come to the house, you know, old people, you're not to run in front of them or bother them. You stay away until called. Every visitor was always given something to eat, nothing special, just some of what we had. No matter if we only had a little. Also if you see old people that need help, always give them a hand. Even if you have only a little, give them some of it. You won't die if you do your share. It has always been told that if you give things and take them back, you will have bad luck. Your house might burn. You mustn't take it back or even use it even if they offer it back to you. If you see ugly people or people with

something wrong with them, "Don't make fun of them!"
It is bad luck. I see now that children are not being
trained about that any more.

Before the people had clothes, there were certain
plants which, if you eat, they said that you would get
hair all over. They preached this to you. It was a big
sin to be hairy, it was a sign you were being punished.
That is what they used to preach. Of course, I know that
if you are part Mexican or 'Mericain,* you are hairy more.

My grandmother told me about what they did to girls
as they were about to become women. But I'm not that
old! They had already stopped doing it when I became
a woman. Grandma told me they dug a hole, filled it
with warm sand and kept the girl in there. They tattooed
her all around her mouth and chin. They would sing
about food and see if she would get hungry; to see if she
could stand hunger. She wasn't allowed to eat. They
danced around the top of the hole. A week, I think they
kept her there, I'm not sure. They didn't want the girls
to get wrinkled early or to get grey, but to have good
health and good babies. This helped them. I would have
gone through with it if they had asked me. I believe in
it, but they didn't ask me. I don't get sick much but I am
grey and wrinkled.

My grandmother was only at the Mission under a
priest for a little while and the Indians still did it then
away from the Mission. My grandmother was tattooed
all over her face but not my mother or father. My husband came from the mountains around Campo and he
had a tattoo on his forehead. It was real pretty, blue-
green and real round, like the moon and about the size
of a half dollar. He had been through the boys' cere-

*Delfina's word.

mony. The people who stayed always in the mountains did these things longer than the people who lived closer to the coast.

They used to take the boys to a special place where they taught the way to be good men. They sang special songs at these ceremonies and tested the boys too, like they used to do for the girls. Some things were done differently for the boys. They made a tu·nak [a hole in their nose] with a sharp stick. It was to make the boys clean. The men used to put a stick or a shell through the hole, or a button of some sort. They used things that we don't know now. My grandfather had a hole through his nose. He was never taught at the Mission, even though he had always lived in the valley until he was old. Some men and women put holes in their ears for ornaments but no one in my family ever did that. The tu·nak was not for ornament.

The reason a young man or woman was tattooed or the man had his nose pierced was that it was needed when you died. It helps you to go on the straight road. If you don't have it, you might turn into a stink bug with its end up in the air and you can't get near the straight road when you die. Now as I am getting old, I wonder if I should have a few lines tattooed so I won't have that happen to me when I die.

Another thing we were told was to get red ants and let them sting us. It would help keep us from getting sick, or if we were sick, it would take the sickness out. Ants were supposed to be good because they ate all sorts of plants and got the goodness out of them. Then the ant put the goodness into you when it stung you. I tried it once and the ant bites made me sicker.

Many stories were told us all the time. The stories used to tell how people are and what to expect from other people in the way of behavior. I only remember a few of the stories now; it is so long since I have heard them.

In the beginning of time lots of wild animals were like people and could talk. There was a coyote and he was a bad man. He was always trying to deceive people and do things that he shouldn't. Then there were these two beautiful girls who were crows and who lived in a tree. An old woman was taking care of the girls and guarding them. But she went to sleep because she was too old. The coyote sneaked up and tried to climb the tree but he couldn't. Then he jumped and jumped but he couldn't reach the girls. The girls couldn't go to sleep because coyote made so much noise jumping. So the girls flew up into the sky and coyote was chasing them and crying and begging them to take him along. The younger girl asked, "Why can't we take him with us?" The older one said, "No, he can't fly." The younger replied, "Why can't we throw something down and pull him up so he can go with us?" But the older one said, "He's too bad; he would eat us." The younger girl must have been falling in love with him because she felt sorry for him. She finally threw down the end of a rope and coyote began climbing up into the sky toward the crow girls. As coyote climbed, he began talking about how he wanted to grab that girl. The older sister got upset and then mad as she heard him talking. She said, "Let's get rid of him. He'll hurt us. He's too different." The older sister cut the rope so that coyote fell and died.

This story explains how we have to watch men—there are some good and some bad men. We knew that these stories were told to teach us how to behave and what to expect. The old people did not have to tell us what the story explained at the end of the story, but I am saying what it meant to us.

Things like that I was told by my grandmother. I still live by the old rules and I've never been sick. I stay away from my daughters when they are pregnant too. When a lady is pregnant, she must not look at anything that is bad, or even see a fox or a snake. You must not look at anything like that or it will mark the baby. You try not to see anything when you are pregnant. The oldtimers would not let a pregnant woman or a menstruating woman go into a garden. She had to stay by herself and not bother anything. She could not gather wild greens, or wash and do things like that. She could not go near sick persons or garden plants without hurting them.

Grandmothers taught the girls that when they were pregnant they must not eat too much or the baby will be born big or have some kind of trouble. They can eat anything they want unless it makes them sick, except they must stay away from salt. Women are weak nowadays. Long time ago, they just kept on doing regular work, they went out and gathered food and whatever was needed, even heavy things, and it didn't hurt them. They just had to be careful not to see bad things.

In the real old days, grandmothers taught these things about life at the time of a girl's initiation ceremony, when she was about to become a woman. Nobody just talked about these things ever. It was all in the songs and myths that belonged to the ceremony. All that a girl needed to know to be a good wife, and how to have babies and to take care of them was learned at the ceremony, at the time when a girl became a woman. We were taught about food and herbs and how to make things by our mothers and grandmothers all the time. But only at the ceremony for girls was the proper time to teach the special things women had to know. Nobody just talked about those things, it was all in the songs.

But I'm not that old, they had already stopped having the ceremonies before I became a woman, so I didn't know these things until later. Some of the other girls had the same trouble I did after I was married. No one told me anything. I knew something was wrong with me but I didn't know what. Food was becoming hard to find then and we had to go a long way to find enough greens. My husband was away hunting meat. Sometimes the men were gone for several days before they found anything. One day I was a long way from Ha-a looking for greens. I had a terrible pain. I started walking back home but I had to stop and rest when the pain was too much. Then the baby came, I couldn't walk any more, and I didn't know what to do. Finally an uncle came out looking for me when I didn't return. My grandmother had not realized my time was so close or she would not have let me go so far alone. They carried me back but I lost the baby. My grandmother took care of me so I recovered. Then she taught me all these things about what to do and how to take care of babies.*

*The European emphasis in the 18th and 19th centuries was on saving the souls of tribal peoples by changing their "religion," forbidding "heathen" ceremonies and destroying the worship of "false gods" and "demons." Europeans did not realize that they were destroying anything more than "false religion" in its narrowest sense. They did not know that they were destroying the total educational, moral, and ethical sys-

After that, I had my babies by myself. I didn't have any help from anybody. My grandmother lived near us but she knew that now I knew what to do, so she never helped me. I did what I had been taught. I used xaʔaˑ nayul [Trichostema Parishii Vasey, mint family] or kʷaˑs [Rhus Laurina Nutt., sumac] to bathe in and I drank a little k ʷaˑs tea also.

I dug a little place and built a hot fire and got hot ashes. I put something, bark or cloth, over the ashes and put the baby in it to keep the baby warm.

So that the navel will heal quickly and come off in three days, I took two rounds of cord and tied it, and then put a clean rag on it. I burned a hot fire outside our hut to get hot dirt to wrap in a cloth. I put this on the navel and changed it all night and day to keep it warm till the navel healed. To keep the navel from getting infected, I burned cow hide, or any kind of skin, till crisp, then ground it. I put this powder on the navel. I did this and no infection started in my babies. Some women didn't know this and if infection started, I would help them to stop it this way.

When each baby was new born, I bathed it in elderberry blossom or willow bark tea. Then after I had washed the baby's face with elderberry blossom tea, I burned some honey real brown, then put water with it

---

tem, which was frequently completely integrated into the "religious" ceremonies. I have been conscious for some years that many of the presently old women of this region had lost their first child, that they were unwilling to discuss the event, and also that none of the living old women had gone through the initiation ceremony. This was a very painful subject for Delfina but she felt that she must tell the whole story so that its truth could be judged. Remember also, she was about 13 years old when this occurred.

and cleaned the baby's face all over. This takes any stuff [scale?] off the baby's face. The afterbirth is buried in the floor of the house.

Some people are not careful and they eat right away and then the mother nurses the baby and it gets infected. The mother must wait a while to eat, then first eat atole. Next, the mother eats lots of vegetables and drinks lots of herb or mint teas. Never drink water! Never eat beans when nursing a baby, it will ruin the baby.

I did all this myself. When my children were older, if they got sick, I used herbs. That is all I used and my children got well again. There are herbs for stomach pains, colds, tooth aches, and everything that the Indians knew. There is a real good one to stop bleeding right away from a bad cut. There is another good one for bad burns and to stop infection. If a woman drinks lots and lots of xaʔaˑ nayul [Trichostema Parishii Vasey, mint family] she can keep from having babies, but there is another herb, even better, that the Indians used to use to keep from having babies every year. They are hard to find now because we can't go everywhere to look for them any more.

I named all of my children myself. I didn't know anything about baptizing them then; I just went ahead like the Indians did and gave them names. When my oldest child was a year or two old, they had a party to welcome him to the group. Everybody got together and they built a big ramada for me and they brought their food together. We had a big fire. I had an uncle that lead the singing and dancing. He led a big xaˑṭuˑp iˑmac [fire dance]. They circled around the fire hand in hand and following each other, and jumping with both feet and singing. They were glad because they would have more Indians, another baby added to the group. All the people brought presents for the baby—baskets, ollas, food, mud dolls, or bow and arrows and different things, whatever was right to start the child. Sometimes they also brought tiny things like the real ones, tiny ollas and baskets and bow and arrows. The child was given its name at the party.

By the time my second child who lived was old enough, we didn't have parties for the new children any more. I don't know why, maybe it was too hard to get enough to eat. I'm just telling what happened to me, what I know.

My grandmother told me that a long time ago it was different. The people used to be stronger, she said. They did not have to use hardly anything to eat when they had those dances. They made a juice from the willow, like a Kool-aid, not fermented. They would drink that and dance for days, she told me. They were naked too, a long time ago, things were so bad and they had so little.

The fire dance was religious; they danced all night, till the sun came up. The songs that go with it have to be sung in the right order, from early evening until dawn. There is a song for each time of the night and as the sun is rising. It was danced at the death of a person and also to welcome a new child. My grandfather said they used to have dances for going hunting, to bring blessings on the hunt; but I never saw one. The dance for a child brings good luck and blessing to the child and to all the people who dance. That is what they did for me and my first child to live. They might have done more before, but they don't even do this now.

When a baby or very young child died, it was just buried near the house of the mother. They did not have any ceremony or any memorial services for them. The child was too young to know anyone, or to have any memories or ties to hold his spirit to this place. Before the Mexicans came, my grandmother told me the people used to burn the dead instead of bury them.

There were many other rules and things we were taught and believed. There were rules so that each one knew what to do all the time. I never killed a snake because that would make them mad and then they might kill a friend or relative. Also, if you kill one snake, more will come back. Snakes and spiders are much alike, both bad. A spider is real bad like a snake, if it bites a person then they can never go into someone else's garden. A scorpion is bad too, but not as bad. You can kill a snake or a spider who has bitten you.

Some years snakes are kind of mean and you have to keep your eyes open all the time. If bitten by a snake, you must get a ʔəwiˑ kusəyaˑy [rattlesnake witch doctor] and get him in a hurry. No one must touch the bitten person but an old lady. There is nothing anyone else can do to help him, only an old woman or the "snake man" can help him. When a person is first bitten, he is kept by himself, only an old person can take care of him. They won't let him near anyone else.

Aurelio, my oldest son, was bitten when he was three or four years old. My husband couldn't get the snake man in a hurry, he was someplace else. My husband said we would have to try something or Aurelio would die. So he cut a cross on his leg and sucked out the poison. He also tied a rag around his thigh to keep the poison from spreading. Then he gave the boy some tea made of mat əyi·w [Euphorbia sp.; several members of this genus have been identified by Diegueño informants for use as a poultice and a tea for insect and snake bites. The name means "earth's eyes," descriptive of the leaves which lay close to the earth.] I had to get away and stay away from Aurelio because I was pregnant at the time. He would be sure to die if I stayed around. I had a little hut to one side that I stayed in. A pregnant or menstruating woman mustn't be around a person who was bitten by a snake. It will hurt the person who was bitten and make them real sick. For one month Aurelio had to stay away from me. I had that old lady, Loretta Quihas of Neji, taking care of him. Aurelio still can't go in anyone else's garden unless he plants it for them. If anyone knows about it, they will run him out.

If a person who has been bitten by a snake goes into my garden, or where I am planting, all the plants will die, or else produce no fruit, or just dry up. The bitten person may plant and work his own garden and sometimes it will be better than ever because he was bitten. Aurelio always has a good garden now. The same thing happens with a black widow spider bite.

[Mrs. Robertson : My grandfather used to get real mad and chase a man who had been bitten whenever he tried to take a shortcut through my grandfather's garden. My great uncle was a snake man. All the snake men had dreamed these dreams and learned these powers. My great uncle claimed that the snakes would come and talk to him and tell him when they were going to bite someone. Then he could be already on his way to the person. When halfway there, he hollered three times and the person laying there could hear him. When Victor was bitten, we children were shooed away, but I stood on the hill and watched. He lay there and the old people covered him up. Then my great uncle came and danced around him and he got well. The bitten person was not allowed to eat anything. In those times everybody believed in the power of the snake man and they got well.]

There is a little animal with reddish fur and stripes [not a squirrel or chipmunk] and when it comes near your house and calls, it means bad luck. When a fox comes near your house and makes a noise, you know a relative is going to die. Also there is a little owl that lives in the ground, kəswanʸ, and he comes to tell you the news when someone, generally a relative has died. If a xatəpaˑ (coyote) howls, it means the same thing. The owl and the coyote are just messengers; the person is already dead. But if you heard an owl outside trying to talk, don't go out to see it because if you do something will happen to you too. A coyote is the worst one, though. But if you stay inside, you can avoid the bad luck yourself. It is bad luck to kill a snake, you should just get out of the way. If you look in a spring to see yourself, a black shadow could cause you to die.

There were bad spirits in certain springs of water. If you drink from them, you get sick and die. There is such a spring above Neji, also one near Ha-a. Sometimes a witch would put a special stone in a spring and you would hear voices and see things if you drank from it. If you stayed around, you would not be able to think any more. All the Indians were scared of the bad springs.

We were always scared of kustəyay [witches] in those days. I don't know how they learned those things. You become a witch if you have that dream. There were two kinds: a good dream made you a good healer to help and heal people, and a bad one made you make people sick. You always knew when a witch came around and you had better do something right quick or he would do something bad to you. You don't ever argue with old people or turn your back to someone at a ceremony. They knew how to put something in you to make you sick and die unless you quickly got a good healer with more power to take it out. They didn't have to touch you to put something in you.

Some witch doctors were more powerful than others. They got their power from their dreams. After a young man had his first dream and received his power, he would work with an older man for a while. There would be special witch doctor ceremonies and dances. They used the toloache [Jimson weed, Datura meteloides]. Witch doctors would come from all over. They would try to use their power on each other to see who was the strongest. They would not let ordinary people come near them to watch. The last xu·ɬuy [witch doctor dance and contest] was held near Campo about twenty-five or thirty years ago. It was on cu·ma· (Tecate Mountain), a special place used only by the witch doctors, high on the mountain. There was another mountain far north in the Cahuilla territory that the doctors went to sometimes. An old relative told me that the doctors were so powerful in the old days that they split the mountain down one side and white sand comes down that side now.

The grownups could always tell if a person was a witch. He didn't say anything. A good one, after he had

dreamed and received his power would go off and fast
and dance by himself. Then he would quietly start healing anyone who happened near. That way people
learned who could heal. Anyone who did that, never
thought of themselves any more, only of the people who
needed help. They were very good people. Even herb
women became like that too, always helping, never
thinking of themselves. No witch could harm such good
people.

The ones who dreamed bad power, if you did something they didn't like, they would make you sick. If they
stopped by your house and you didn't give them enough
food or made them angry in any way, then you would
get sick after they left. I was never bad to anyone so they
never bothered me. I've never been mean so I didn't have
to worry. But we don't have witches anymore, nobody
dreams power like that. There are a couple of men from
somewhere else who pretend to have the power but most
of the Indians know they are fakes.

When I was young I had an old relative. He was real
old and almost ready to die so I asked him how he became a healer. He told me he was the youngest of five
brothers and each one had been given some power or
other in their dreams. Then one night when they were
all asleep, he was chosen. He was awakened by a man in
a fireball drawn by two horses and with a dog, all golden
and shiny.* The man told him not to be afraid but to
come with him. He wasn't afraid and the man took him
to Signal Mountain near the Colorado River. They flew

*Note the European elements in this story, the fireball (a chariot), horses,
the three offers, the gold and then the Indian element of healing by
sucking.

through the air and then went inside the mountain. It was all golden inside. There were some men there and they told him not to be afraid. They asked him if he was hungry and to go ahead and eat the food there. He was suddenly real hungry but he looked and there were people there all skinned and hanging like meat. He said, "no." He didn't want to eat. Then they showed him some piles of gold. Did he want that? He could have all he wanted, and he would always be rich, but he said, "no." Then there was a poor old sick woman sitting there full of sores. He wanted her to get well. They asked him if they should kill her to get rid of her and the sores. He said, "no." He wanted her to get well. They said, "Well, will you go over there and put your mouth on her sores three times and spit it out." So he did and her sores went away. They said, "All right, you don't belong here. You go back and start healing everybody. You'll never get anything." They flew him back in the golden thing and then he woke up. The next day an old lady came by, all full of sores like in the dream. So he put his mouth on them and they got well. He was always able to heal people after that. The only way to heal someone who was sick because of a bad witch was to get a healer who had more power than the bad witch. This old relative was very powerful. He told me, "Don't ever take anything you are offered in your dreams because the riches and food won't be good." After he had his dream he told me that he had to go up onto a special mountain [Mt. Tecate] for five days. His healing songs came to him during that time.

He was real old when he told me and he only told me because he knew he would die soon anyhow. But he knew that he would die before the year was out because he had told how he became a healer. He did.

My grandmother told me she was under the priests at the mission for a little while when she was young. She never told me anything about how they had to work at the San Diego Mission. The Indians around Ha-a and Neji talked about the ones that used to be down there. The Indians did not like them because they had to work too hard for the priests. The women had to make a lot of things. They wove blankets and bedding, made ollas, and learned how to make mud houses. The Indians either learned and did it or they were punished. The Indians left when ever they could. They said the priests were all bad because they made the Indians work.*

I just heard about priests; I never saw one. I don't know what the old Indians or the priests used to do or say about God. Nobody ever told me anything about God, that I can remember. But I thank God all the time, especially for plants. One year plants and things are good, lots to gather, and I thank God and then ask Him to give us a chance next year. I still do that myself now. As I gather, I thank God. All that I wish for now is for my children. I ask God to help find work for them so they can feed their families and themselves. I ask for myself to get older so I can see my grandchildren grow up. I ask that my grandchildren can go to school and learn the new way to live and work. Only good health and work for us all is what I ask. I don't know anything about priests or church, this is just talking to God myself and asking God's help.

*The Indian memory of the Dominican missionary period in northern Baja California is still quite vivid among the older people.

just
w   ace
w   ally
ca   ned
th   ith
th   ere
al   ow
ar   ole
ar   old
us   nan
wl

My mother and father separated after we had been there a while. I don't know why. My father left and never came back. Then my mother said I had to get married so that there would be a man to hunt for us. My grandfather and grandmother were getting too old and needed someone to hunt for them. Sebastian Osun was the man they picked for me. He had already asked my grandfather and grandmother for me. My father had already left and then my mother left. They told me to marry Sebastian because I needed a good man to take care of me. Sebastian was the kʷaˑypaˑy (captain) at Ha-a. He was from up near Campo but he was such a good man that people there had asked him to be kʷaˑypaˑy.

The kʷaˑypaˑy took care of the Indians in a village. When there was trouble in the village, he acted as judge. If there was trouble with some one outside the village, he had to fix it. He called the villagers together for ceremonies, deaths, and other things. He didn't tell anybody what to do, but he had to help anybody who needed help. Sebastian had a relative already in the village who was the kucəkʷaˑr (speaker). He had to talk to the people during ceremonies and explain how things had always been done. Sebastian wasn't much of a talker but he was a real good man.

In those days a girl stayed with her parents or grandparents until a man asked them for her. He would bring food to her parents as presents and then ask for the girl. If they accepted the food and the man, then they gave a party for everyone to announce the marriage. Sometimes they had a dance as well as the feast. As food got harder to find, they stopped having the feasts.

So I had to marry this man. He was the father of my children. He was born somewhere near Old Campo, an Indian reservation near Cameron Corners. He was related to Mrs. Robertson through an uncle. Sebastian's uncle was Alessiano Manteca, a kun^{y}i·^{ły} (Quinnich), who was also the uncle of Rosalie's grandmother. So Rosalie's grandmother and my husband were first cousins. Lots of people moved around then. If some rancher settled near their place, they had to go elsewhere. A lot of them who wanted to live the only way they knew went south where there were not so many settlers and they ended up around Ha-a. Some scattered into other places down there too. There were Indians in all the little valleys where there was water then. Most have died now.

When I married, everything was all right; Sebastian was a good man. He worked hard. Sometimes he raised cattle and sold it. Sometimes he went all over looking for work. He worked different places, whatever someone wanted done. That's how we lived over there. He was a good man. He was real good to me! He took care of the children. He took care of my grandfather and grandmother. Oh! He was good! I had no trouble with him. He took care of me and we always had some food to eat.

My husband used to hunt a lot. He was a good hunter for deer. We would use some of the meat and some we would trade for sugar and flour and coffee. Sometimes I would gather medicines and herbs and trade them off for food. We had enough food then with both of us bringing in everything we could find.

There was one Mexican family that lived near there with an old woman who wanted herbs. They would bring food and trade it for my herbs. There was no store. We didn't understand money then; the only thing we did was to trade for food or cloth. Sometimes we were lucky and found a lot of honey and we could trade that for a cow or a calf to raise. I hunted for greens and roots and seeds as I had been taught. We used some honey for ourselves. We used everything available for food. We hunted təkši· (gophers). The place where I live now is called by the Indians təkši· nʸəwa· (gopher's house) because there are so many gophers there.

We traded to get cloth and made our own clothes. We never wore store shoes. My grandfather used to make shoes from deer or cow hides. He would get some hide from someone who was butchering. He also used to make sandals of yucca. The word "kʷak" means deer but it also means meat in general, depending upon how it is used.

Our family went to the beach down below Ensenada and to Rosarito Beach when we couldn't get to the San Diego beaches any more. Lots of Indians went there every year also to fish and to look for abalone. The women dug clams and looked for all different kinds of shell food and plants that grew near the sea while the men fished. We opened up the shells and dried the meat on the rocks in the sun. The children helped their mothers, even with pounding abalone. Some Indians made necklaces of shells, but we didn't. We always needed more food; we were poor and never had time for necklaces. I don't know how to make those things now, only how to find food. Some years we went to San Felipe on the Gulf for fish. Whenever the weather was hot, we were at the ocean or at San Felipe, every year. We would camp on the beach or just back of the beach near a spring. Sometimes others were already camping near the springs so we had to camp farther away and carry water to our camp. When we were at the ocean we always caught lots of the little fishes that came in at night and dried them to use in the winter.

We spent the winters at Ha-a. There were acorns there. We gathered them right near where we lived. Then when things began to grow in the spring, Sebastian and the other Indian men would go to the low mountains on the edge of the desert for about a week. They gathered agave, dug large pits, and roasted the agave in the pits. They used wooden shovels that they had made from mesquite.

The agave gathering and roasting was men's work. Hunting game for meat and hunting for bees and honey were men's jobs also. The women hunted for wild greens, seeds, and fruit. The whole family helped with gathering acorns and pine nuts.

Each fall our family, when I was young and later when I was married, went to the Sierra Juarez to gather pine nuts. There are some rancherias there where some Indians live all the time. There was lots of water in those mountains so we always camped near water. Sometimes we would camp near Tres Pozos or Mesquite or Jasay.

Sometimes after we had a lot of pine nuts, we would go down to Santa Catarina. We used some of the nuts to trade for other things that we needed. Once we went on down to Valle de Trinidad, beyond Santa Catarina, and met lots of Kiliwa. Another time we went to Arroyo Leon to trade and to attend a fiesta with the Indians there. Sometimes we went to Jamul to a kəruk.

We went down into the desert near Mexicali every year to get mesquite beans. Mesquite beans make a good flour for bread. They can also be soaked in water to make a good cool drink. No matter what we were gathering, I always looked for herbs and other greens where ever we were.

There used to be lots of Indians in all the little valleys down there when I was a young woman. There aren't any Indians there now except in a few places. There are still some about twenty miles below Neji at two little places called San Pedro and San Pablo. There are a few families at Los Coches. There was a village near Guadalupe that we used to visit. There were even some Indian rancherias near Tia Juana when I was a young woman. We used to visit them with Sebastian. There are still some Indians at San José, east of Tecate. There used to be a village in the big white rock mountain above San José. Only one man lives there now. There were five, ten, or sometimes more Indian houses in all these places and other places where I only know the Indian name.

As I roamed the mountains looking for food, I have seen lots of grinding holes in the rocks everywhere. There must have always been Indians all over the country.

We would go to all these villages if there was going to be a dance, a kəruk, or a fiesta. We would take the children with us and go to the ceremony and dance one or two days and then return to Ha-a. Sometimes we went far. We had to walk for several days; we had nothing to ride. I liked to dance and be together with all the people that we knew from all over. The Indians used to dance for a death, one year after a death, a wedding, a year after a baby's birth, and sometimes just for fun, especially if we had had a good year and had lots of food stored for the winter.

I think my biggest boy was about 11 years old when Sebastian died. My husband was sick and then one day he had to get up and go after something—I don't remember—and the horse threw him and it killed him. Oh, he was going to hunt for some honey to trade for food because we were out of food.

When my husband died, I cut my hair off, all real short. That was the Indian way. Then let it grow out again. Then we had to burn everything that belonged to him, or that he used, or made while he was alive. Only when everything is burned can his spirit go into the next world and not have to keep coming back after his things. Then he will have the things he needs in the next world too. It has to be done mostly because his spirit can't leave, but is troubled and keeps returning for anything you keep. When my husband died, we had to burn everything, the grass house and everything. Also we built a fire in the open and put lots of sage on top. Then we leaned over it and got all smoked. This purifies you after a death in the family. It helps you so that you don't remember and grieve too much. I did it after my husband died. Burning everything and purifying with sage was the law then. It would show that you didn't like him and didn't want him to rest if you kept anything.

I knew how to build another house, so I built one after the funeral was over. Mourning is for one year after a death. Then I had a ta·kay [the anniversary service held one year after the death of one person—the kəruk is a special ceremony for all the dead]. It was hard to get enough food to feed everybody but he had a lot of family and relatives and they all helped at the ta·kay. For his image, we used some clothes like he used to wear. The person who dances holding his clothes represents him dancing for the last time.*

After my husband died, I had a hard time getting enough food for my children. Things got pretty bad. I went out on my own and gathered food. I had been taught all these things about how to gather and prepare wild food. I went out and hunted for wild greens and honey. I took Aurelio and we hunted with his bow and arrows and his rabbit stick. Sometimes we found things. Lots of times we did not and we went hungry. I had to beg for food from neighboring Indians and ranchers. Some neighbors helped me sometimes. I went hungry and my children were hungry. Sometimes for two or three days we found nothing. I didn't have anybody to help me any more, I just went here and there looking for food.

I finally had to sell Aurelio to a Mexican to get food.

*Most literature describes the anniversary service held one year after a death as a "karuk." This Southern Diegueño distinction between the karuk and the ta·kay became clear only after attending several services and noting one image or several images present and questioning informants carefully.

He and his wife said they wanted a son because they had no children.

Aurelio was about twelve years old. They didn't treat him like a son. They were mean and made him work like a man all the time and even beat him. The food they gave us to pay for him lasted a month. Aurelio stayed with that man and his wife for three or four years, working like a man for him. He finally ran away because he wasn't going to be beaten again. Aurelio couldn't stay with us then because that man came looking for him and would take him back and beat him some more. Aurelio worked here and there on various ranches and sometimes hunted. He brought me a little food for the smaller children whenever he could. While he was a boy, he was never paid in money for any work, just food.

Santos was just a baby when his father died. One time we walked into Tecate looking for food and to hide from that man, just baby Santos, and the girls, and me. We went back because we couldn't get anything to eat in Tecate. Out in the mountains at least we had what wild food we could find.

When Lupe was eleven or twelve, a man wanted her. Bernardo Mata wanted her. His mother said they would give us food from time to time if I would give Lupe to be the wife for this man. So he married her and she went to live in Neji with the Mishquishes. That man beat her and his family made her work real hard all the time. Finally I went and took Lupe back because they were mistreating her so. I had to leave Lupe's son with the Mishquish grandparents or they wouldn't let Lupe go. They took care of him even though they were mean to his mother.

The terrible things I went through trying to keep my children together and fed, I can't begin to tell. Then, I didn't succeed after all. I feel like crying when I think of that time. My children were hungry and cold so many times.

I tried to live with several different men, each one said he would take care of me but each time it was always the same. I did all the cooking, washing, ironing, and everything, all the work I had always done, but it wasn't enough. I had to cut firewood and stack it. I had to clear land and cut fence posts. I had to work like a man as well as the house and garden work, hard, heavy work. If I didn't do enough to suit him, he would beat me. I have been black and blue all over so many times because there was still more work to do. Even with all that, each man would get mad about feeding my children and beat me for that. When the man would not let me feed my children, I would have to find some one else to work for. I was beat so many times. It is hard work to dig fence holes and put up fences and everything, but I had to try to get enough food for my children.

I got so I couldn't stand any more, I couldn't get enough food for them. Then Juan Tampo promised he would be good to Lupe if I let him marry her. So he married her, but he just started beating her too. He just wanted someone to do his work for him. When he almost killed her one time, I took her away from him. Lupe went through a hard time before I took her back.

I had to give Eugenia, when she was real little, to her Godmother, Matilda Osun. They had to take her because I couldn't get enough food. Matilda had a good man for her husband. Eugenia didn't get along with them so later I took her back.

After several years, I married Lupe de la Rosa, a Mexican from Los Angeles. I stayed with him quite a few years. When I first went with him to Tecate, Lupe had just married Juan Tampo, and Lola had been asked for by Domingo Calles of Ha-a, and Eugenia was with Matilda. I only had Santos with me. Then I had to take Lupe back; then Domingo started drinking and kicking Lola around so she fled back. They all thought maybe we could all be together again because this Mexican man seemed all right. But he didn't want my children around. He didn't want to share the food with them even though I was working too. I was washing and ironing for other women around Tecate. My Mexican man began fighting with me more and more.

Santos was finally old enough to leave and to begin to work here and there. He began working while he was still a little boy. He just worked around for food where ever he could, from place to place. Now he is a good steady ranch worker and works all the time. He has never married.

Finally Matilda Osun's son Alfonso Perez wanted to marry Lola. They have three children. He treats her pretty good and brings food for her family. Then Eugenia married a Mexican. She has children by this Mexican and he treats her pretty good and always brings food to his family. They live in Mexicali. Aurelio has never married either, but he is a good steady ranch worker and always has work now.

I left Lupe de la Rosa finally because he didn't even want any of my children to come to see me even when they were all on their own. He'd get real mean and beat me every time one of my family came around. We didn't get along. So I left him and stayed by myself for a long time. Sometime one or another of my daughters stayed with me for a while.

About four years ago I went to live with this man who is part Yaqui from Sonora. He has a little place near La Rumerosa. My marriage with him isn't very good any more. He takes everything I have and sells it for money to get liquor. He shoots at me or right along side me lots of times. We fight all the time. Sometimes he goes off for months and I'm by myself. I earn my food now the way I did in Tecate, by doing washing and ironing for the ladies on the ranches around here. There are too many people all through the mountains now for Indians to live by hunting and gathering the wild food the way we could when Sebastian Osun was alive and I was young. The Indians are having a hard time unless they can get jobs now. But with more families, there is more washing and ironing than there used to be. Like the other Indians, I keep a few goats and chickens now, too. I try to work real hard. I chop and gather firewood and cut fenceposts and put up fences still.

There is a white man that comes from someplace near Los Angeles. He has some of the people going all over looking for gold. He takes a machine over there and whatever we find, we bring to his machine when he comes and he grinds it. Some other women work with me too. We carry the rocks on our back from where we dig them to his machine. He pays us with food. He wouldn't pay money! It's real hard work. My husband doesn't do that, it's hard work!

I have my sister and brother over here and one of these days if my husband kicks me again, I'm going to try to come back where I was born. He gets mad when my children visit me too. I'm not getting any younger and it is getting harder and harder to do this work. I don't think I can go on much more. I would like to come back to where I was born for good if I could. I would do anything to work to make a living.

I don't have many relations still alive, just a few and my children and grandchildren are all that are left.

After my father left one day and never came back, he went back to Jamul and Campo. I heard he married again. Later I heard he was hit on the head and he was put away somewhere after somebody beat him up. He died in 1963 in Patton Hospital in California. He had been too old to take care of himself for a long time. They let me come to his funeral at Jamul.

My mother left just before I was married because I was a woman and could care for myself. She later married again, a man named Domingo Cuero of Campo Reservation. He was no relation to my father unless way way back. By this man she had two sons named Jose and Alberto Cuero and two girls, Viola and Ruth Cuero. They are living somewhere in El Cajon. Later I heard my mother left Domingo and then she married Mosalino Fernandez. By him, she had one son named Alex Fernandez. I have never seen him but I heard he is living in Phoenix, Arizona, and married to a Pima woman.

I have many relatives in San Diego that I haven't seen for years and I don't know if they are living or dead. There is a lady named Ramona Cuero. There is my cousin Bernardo. He is a cousin that Indians call brother; his father and my father were brothers. Henry Serrano is a cousin. Isabel Thing [deceased January 1966] was a cousin. Isabel Rosales who lives in Jamul is an aunt on my mother's side, my mother's sister. Then there are the Quaha cousins, Ortiz Lopez (Quaha) who died in 1963 was my mother's brother. My mother's half brother, Raymundo Quaha [deceased], was enrolled at Campo Reservation. The other Quaha's left besides Fernando and his family at Ha-a are Laura, enrolled at Campo, and Catalina, Vivian, and Rosalia.

My husband's brother was Matellio Osuna who was enrolled at Manzanita Reservation. Margaret Osuna Brown of Baron Long [deceased 1966] was his sister. She was my children's aunt. Then I have a niece, Mrs. Virginia Domingo in Yuma, Arizona.

I heard that I had two sisters older than I, but they died before I was born. Those are all my relatives that I know, still alive, or who have died just recently.

I had two older girls that died young. Aurelio is my oldest boy. Lupe is the oldest girl alive, then Lola and Eugenia. Santos is the youngest child. So I have five alive. All my children were born at Ha-a by the man I married first, Sebastian Osun, a good man.

My two boys haven't married and don't have any children. I have ten grandchildren. Guadalupe's son, Silvano Mishquish, is the oldest. Then she has three girls, Josephina, Martha, and Lupita. Lola (Maria Dolores) has a girl, Patricia, and two boys, Jose and a new baby. Eugenia has a boy, Jose, and three girls, Delfina, Maria Teresa, and Margarita.

I keep praying to God that before I am too old to work for my living I can come back where I belong and be among the few relatives I still have alive. I pray that something will work out so that my children and grand-

... which some families managed to survive the disintegration of their economy and social structure under the pressures of a stronger, more aggressive civilization. We have seen their struggles, hopes, and fears as they clung to and yet realized the inadequacy of their old life. Strength and resiliency are the characteristics of the individuals who are still fighting to retain some of the good ideals of the old life while searching for survival in a new uncertain future.

These people have managed to maintain their independence during the time that many Indians were forced into dependency upon the federal government. Now they are asking, "Can our grandchildren go to your schools? There is no longer any room for hunters and gatherers! We can no longer teach them how to survive. They must learn the new skills from you!"

Delfiina's reason for telling her story and her main question is yet to be answered: "Is there room for us in America? Can we come home legally?"

Originally printed 1968 by Grant Dahlstrom at the Castle Press, Pasadena, as Volume 12 of the Baja California Travel Series. New edition 1970 by Rubidoux Printing Company, Riverside, California.

AN ACCOUNT
OF THE REST OF HER LIFE

While the first edition of her autobiography was in press, in 1967, Delfina was at Jamul taking care of an invalid Kumeyaay (Ftnt.l) (Diegueño) woman. A social worker came to interview the invalid and she saw Delfina. She asked about Delfina, who she was, where she was born and when. Upon being told that Delfina's birthplace was Jamacha about 1900, the social worker filled out the necessary forms and placed Delfina on Old Age Pension, In a few months, the Jamul woman's health improved and she no longer needed care. Delfina asked if there was room for her to live at Mrs. Rosalie P. Robertson's house on Campo Reservation. At her reservation house, Mrs. Robertson fulfilled one of the traditional functions of a *Kwaaypaay* (village leader) and of the *Kuchutt Kwataay* (tribal leader) from whom she was descended. To provide living places and care for the elderly, she had added several rooms to her reservation house and built several small cabins beside it. Because Mrs. Robertson's husband worked in San Diego, the Robertsons also rented a house in El Cajon and had a room there for the elders to sleep when they had to come to town. At the reservation house, Rosalie had a phone installed, but most of the elders did not like to use it, so Rosalie or her brothers went by the

Campo place weekly to check on the needs and condition of the elders. Over the years, at any one time, she had as many as five extremely elderly Kumeyaay there. She brought them regularly to town for their shopping and health needs. Rosalie always took them to all the ceremonies in the county, to Quechan and Cocopa ceremonies near Yuma, and to the villages in La Frontera, Baja California. To provide the elderly with their favorite meats, several of the teen age boys on the reservation regularly hunted rabbits, quail, and other small game, and occasionally deer. All the game was given to the elders, according to the traditional beliefs that young men and their families could not eat the game they caught, but must give it to the chief for distribution to the elders and the sick.

Delfina did not want to live in the Robertson house in town. She wanted to be out where she could plant a small garden and keep some chickens. She wanted to enjoy roaming the hills for medicinal and food plants. She used the bedrock mortars on the slope above the reservation house for grinding the acorns she gathered. She pecked small pits ("cupules" as the archaeologists call them) on the bedrock and used the holes for holding the nuts as she cracked them. In her walks over the reservation hills, she located wild roses, manzanita, wild plums and gathered their fruits. She made teas from the flesh of the fruits and ground the seed for food. She located several varieties of opuntia and cholla cactus from which she took the fruit and young new pads every year. She preferred the original foods and harvested everything she could find. She cooked and dried the greens and some roots for later use. She gathered and stored medicinal plants for herself and the other elders.

Among the aged at Mrs. Robertson's was an elderly Kumeyaay man who became Delfina's special friend.

They attended all the ceremonies, fiestas and other special events, sitting together, holding hands, conversing happily. They participated in traditional dancing. They gathered acorns together and Delfina cooked for him. Delfina ground the acorns, leached the flour, and cooked in the older traditional ways - by flavoring her *shawii* (acorn meal) with a variety of other finely ground seeds and plants. Some *shawii* became like rich desserts with nut-like flavors, such as almond flavor from the wild plum seeds (but no sugar was ever used). Other times she flavored the *shawii* with wild onions, sage, or spicy seeds to make an accompaniment for meat and vegetable dishes. She gathered fresh greens, roots, cactus and fruits and used traditional foods as much as possible. She was an excellent cook. Once, she commented that in the past most foods were boiled and a few roasted. I enjoyed her meals every time I visited her on Campo. She was always telling us stories about the past and old traditional children's tales. Or we would all be laughing and joking over some recent event. She was a happy person and this was a very happy time for her.

Delfina's two sons came to Campo to live and the necessary identity papers were provided for them. Mrs. Robertson and I took them to social security and draft registration. Delfina's two daughters remained in Baja California but came regularly to visit, bringing their children. With the royalties from the first paperback edition of her autobiography, published by Malki Museum, Mrs. Robertson and I bought the gifts which Delfina wished to give her children and grandchildren.

One time, Dawson's Book Store held an authors' party for all the authors of the Baja California Travel Series, so I drove Delfina, Mrs. Rosalie Robertson and Dr. Margaret Langdon to Los Angeles. To avoid heavy traffic, we left San Diego early and ate at a restaurant on the way up. Delfina also enjoyed eating at restaurants. At Dawson's, Delfina was introduced to the guests and other authors. She tried and liked the hors d'oeuvres and different types of food served at the party. She enjoyed meeting and signing everyone's copy of her book. The rest of us who had participated also signed the copies. When we drove home, she told us how much she had enjoyed meeting all the people. She was grateful for their interest in her life and problems.

Within a short while of Delfina's settling at Campo, her son Aurelio (who spoke slightly more English than her son Santos) obtained regular work on a nearby ranch. Often, he and his co-workers went to Tecate, Baja California to buy their supplies and clothing. Several years later, on a payday when they went to Tecate, someone there knifed and killed him for the cash he was carrying. The police never found the murderer.

Another year, Santos was working near Jamul with a Kumeyaay from there. Unfortunately, after work the two men celebrated at a tavern and then began driving home. That is the last that was heard from them for two months. They seemed to have disappeared from the face of the earth. No one knew what happened to them; when inquiries were made at the sheriff's office by Mr. and Mrs. Robertson, deputies there said they knew nothing about the missing men.

About two months later, at her request, Mrs. Robertson's brother, Chris Pinto, stopped by the Campo Reservation house for the weekly contact with the elderly to make certain they were all right. Delfina showed him a letter she had received and asked him to read it. The letter was from the San Diego County Mental Health facility stating that a hearing was to be held to determine if Santos was dangerously schizophrenic and should be incarcerated in Patton State Hospital. If his family had any objections, they should be at the hearing at the Facility at 8:00 A.M. Monday. This was Saturday evening when Chris had stopped by. He quickly had Delfina take her town clothes and come with him down to Mrs. Robertson's house in town. Rosalie called me and alerted me. Then on Sunday afternoon, during visiting hours at the Facility, Rosalie's husband, Carl, and Chris went to see Santos to find out what had happened.

Santos and his Jamul friend had been stopped by the sheriff and taken to the county jail for drunk driving. There they were given physical examinations. The doctor conducting the examination determined that Santos needed a colostomy. We were unable to find out why, or what condition existed, nor how anyone obtained informed consent for the operation. No relatives were notified. Santos was taken to San Diego County Hospital and operated upon. After about four weeks, the hospital sent him to a recuperation facility or nursing home where he was to remain until completely healed.

After another week at the recuperation center, Santos was beginning to feel better. The nurse came into his room to bathe him and change his colostomy bag. He reacted to her cleansing of his genital area, assuming that by touching him there she was making sexual advances. She screamed and fought him - a doctor entered and Santos threw a chair at the doctor. He was finally subdued and then taken to the Mental Health Facility. And the letter was sent to his mother.

Monday at 8:00 A.M., I entered the Facility foyer with Mrs. Robertson and Delfina. We were informed that only family members could attend such hearings. We pointed out that Delfina did not speak English and needed a translator and Mrs. Robertson stated that I was their anthropologist and would be needed to explain the family background and explain the meaning of any technical terms to her for translation. We sat and waited two hours until the hearing for Santos was called.

We entered the hearing chambers. The judge sat at one end of a table; Santos was seated at the other end, and a very grim looking young man and a severe looking older woman sat on the far side. We learned they were the psychiatrists who had separately examined and questioned Santos, and thus had determined that he was dangerously schizophrenic. We were ushered to the near side and introduced. The judge opened the hearing by reading the charges relating to Santos's conduct: that he had attacked the nurse and doctor. Then he read the psychiatrists' findings that he was dangerously schizophrenic. The judge asked Mrs. Robertson to translate his statements. Next, he asked how Santos would plead, guilty or not guilty. At this point, I nudged Rosalie and whispered that she should ask the judge if I could explain the situation; which she did. The judge looked at us with interest and stated that he would be glad if I could explain the events.

I began with Santos' childhood, which resembled Aurelio's. In order to feed her children, his mother had let a Mexican family take him when they promised to treat him as their own son. Instead they had abused, overworked and underfed him. Santos fled, but could not remain with his mother because the Mexican man came searching. Santos then worked all around for various ranchers on both sides of the border - thus, as an adult his command of his own language was that of an eight year old, while his Spanish and English consisted of ability to follow orders on ranches. Also I stated that living this life, he had never had any medical care or been to a doctor in his life. How could he possibly have given any kind of consent to the operation? How could he have had any understanding of what they planned and did to him? How could he understand what our medical doctors and nurses were? I explained that as he was feeling better, his response to the nurse was a culturally appropriate response to a female touch in the genital area. I also pointed out that persons unable to understand the language of a questioner often watched closely to determine, if possible, whether the person wanted a yes or no answer, and answered yes or no, as they thought the questioner wanted. Certainly Santos could not possibly have understood the psychiatric interrogation any more than he understood the medical examination and operation.

The judge looked over at the psychiatrists, and they both nodded and said yes, it could all have been a cultural misunderstanding. What else could they say! Santos was then released to his mother for returning to the reservation. A visiting nurse was sent to teach his mother and Santos how to change his colostomy bag. The other Kumeyaay man had been in jail for drunk driving all that time. His family had not been notified of his whereabouts until Mrs. Robertson called them after Chris learned from Santos that the other man was in jail.

Mrs. Robertson and I were concerned about other Southern California Indians, especially the elderly, caught in such language and cultural difficulties when no one was available to speak for them who could understand and thus stop what would be a railroading process. We both had heard of other Indian cases in the recent past. However, no one at the facility was interested in our concerns at that time.

Santos finally was able to begin working again. Life resumed normality for Delfina and her friends, who continued to go to ceremonies and to enjoy all the traditional ways of the Kumeyaay.

From time to time I worked on ethnobotany with Delfina. After many rainfalls we made repeat visits to places that had remained untouched by modern developers near the coast and in the nearby valleys. Delfina's knowledge of plant foods and medicines was voluminous. A botanist working at Torrey Pines State Park asked if Delfina and her family had ever gathered food at Torrey Pines, and on receiving an affirmative answer, asked me to bring Delfina to identify plants at the park which the Kumeyaay had used. Because nothing could be picked, the botanist and a ranger were to go around with us to the different plant locations and identifications were to be made on the spot by the botanist.

When I brought Delfina, Rosalie Robertson and Dr. Margaret Langdon, the linguist, to the park, we met the Ranger, but the botanist had some emergency and could not be present. Instead she sent two amateur botanists to replace herself. Because plants could not be picked, or sample branches taken, we had to identify plants as we walked around the park. With so many people asking questions, this process became confusing at times, but we did manage some plant-use identification. Delfina stated that when she was young and they harvested nuts at Torrey Pines, there were more pine trees than now. Also that scrub oaks were around the edges of the pines and they also gathered the acorns from them. At the park, we spent most of the day walking over the mesa looking at plants. Delfina exclaimed happily over a small succulent that she had not seen since her youth, declaring that it was extremely tasty eaten raw by itself or added to other raw greens. She spoke to Rosalie who translated to me that she was requesting permission to take one small side shoot that she would be able to root out on Campo. The others heard her request and rapidly walked on out of sight.

Another time I arranged to take Delfina and Mrs. Robertson to the Mission Bay Salt Marsh Preserve (under the care of the University of California at San Diego) where we were met by a caretaker and walked through the marsh identifying plants used by her family in her childhood. She kept looking for one with a small white potato-like root but we were unable to find it, even though we found many other plants that her family had formerly used.

I discovered that Delfina knew which plants would root from a branch or side shoot. She knew which could be increased by root division or transplanted to a new location. She told us that her family told her that before the non-Indians forced them to continually move, but when they could stay on their own land, they would gather seeds from some plants, and cuttings or root sections from others, and plant them near their house in a clearing they had made. Other plants were transplanted to their land. She was slowly moving plants from distant reservation locations to a hillslope near her living quarters.

Delfina also provided some information on Kumeyaay use of fire to manage plant growth, plant diseases, plant parasites and destructive insects. Many areas were burned each year just as the plants began drying, before the fire could spread because most things were still too green. She said they did not allow an accumulation of dried leaves or pine needles and other plant dropping to accumulate. After the trip to Torrey Pines, she commented that there were more weeds, underbrush shrubs, carpets of dried pines needles, and broken branches than she had ever seen when she was young and with her family had gathered food and pine nuts there. She was afraid there was so much fuel that any chance fire would destroy the trees. She said the Indians did not allow that much fuel to accumulate under oak or pine trees because that trash reduced the quality and quantity of the nuts available for harvesting, as well as causing damage or killing the trees by fire and plant diseases. She said that burning every year never let enough fuel accumulate to damage the trees. She also said that she had been told that some pine nuts had been planted to increase the size of the grove.

On another occasion, Dr. Margaret Langdon arranged for Rosalie to bring Delfina to her linguistic class at the University of California, San Diego. Delfina charmed the students with a number of legends and stories. She also sang and showed the students how to dance, dancing with each student. Delfina told Margaret and Rosalie how thoroughly she had enjoyed the visit. Margaret responded that the students had loved having Delfina and had a marvelous time also.

Then in May, 1972, on her weekly visit and supply trip to the elders on Campo Reservation, Mrs. Robertson discovered that Delfina was ill with severe abdominal pains and had not phoned for help as instructed. Rosalie took Delfina to the hospital immediately. Delfina was admitted and the doctors operated and discovered that Delfina had a burst appendix. She remained unconscious for the next two months. The doctors stated they were amazed at her bodily strength and ability to battle the infection from her burst appendix.

They kept her in the contagious ward in a private room hooked up to many tubes (intravenous, nose and catheter) and monitoring machines. I was in Hawaii working on my doctoral program at that time. I took leave of my professors and flew back. Mrs. Robertson and I were allowed to visit her only on Wednesday and Sunday afternoons (contagious ward visiting hours). We were there one Wednesday when she became conscious for the first time. Rosalie explained what had happened and what all the tubes meant and what was being done for her. The doctor felt that she had improved somewhat and had some chance for recovery. However, she was extremely frightened by all the monitoring apparatus, intravenous tubes and strange surroundings. We kept reassuring her and told her we would be back on Sunday. We came, and she still expressed fear of the medical people and their handling of her. There was no one to tell her what they were doing on a daily basis. We had been unable to get permission to come for a short time every afternoon.

The lack of anyone to communicate with her every day, combined with the total strangeness of her surroundings and the tubes was just too much. She became more and more fearful and lapsed back into a coma. At this time, Anna Sandoval, of Sycuan Reservation, aided by bringing Delfina's two daughters from Tecate to visit their mother at the hospital. They arrived near the end of visiting hours and the nurse turned them away. Anna objected to the nurse's refusal to let Delfina's daughters visit her. Finally the hospital administration allowed the daughters to visit Delfina and sit beside her for a while. Before the next week was over, Delfina died. The fear was too much to overcome.

Mrs. Robertson held the clothes burning, wake and funeral, and buried her at Campo. A year after her death, Mrs. Robertson held a memorial service and placed a headstone on the grave, Mrs. Robertson and I paid for the wake and funeral, for the memorial service, and also for the headstone on Delfina's grave in the Campo Reservation Cemetery.

Despite these sad episodes, Delfina's last years were lived in comfort and happiness on Campo Indian Reservation, supported by Old Age Pension. The royalties from the Malki Museum Press paperback edition of her story were used for special items not covered by Old Age Pension or for gifts and cash for her children.

Several years after Delfina's death, Santos died from cancer of the colon. If we had only been informed of the original doctor's findings, perhaps we could have had Santos cared for more effectively.

Problems Remaining

One last problem remains for Delfina's family. Several years after the death of Delfina, her daughters finally decided they would like to return to their mother's home and come back to the United States where all their relatives still live. To date we have not been able to get any help from the United States Consular offices in Baja California. Yet, for years this country has accepted refugees from many countries around the world.

Are we unwilling to allow the return of the children of people who became refugees from the United States itself?

Next, would any of the San Diego Kumeyaay Indians (those mentioned in my 1968 introduction to the *Autobiography*, and others I have since met) have become refugees, if the Catholic Church had honored the trust deed to the lands of Mission San Diego, signed by Abraham Lincoln? The trust deed states "in trust for the religious purposes and uses to which the same have been respectively appropriated." If, instead of leasing Mission San Diego land to non-Indian farmers and eventually selling most of it, the church had used Mission San Diego land as a refuge and haven for dispossessed Indians according to the deed in trust, would refugees from California and the United States be living in Baja California, Mexico? There are still many San Diego County Kumeyaay, refugee descendants, immediate relatives, brothers, sisters and cousins of the people on the Kumeyaay reservations. Some of these people would like to come back. Others would like the freedom of border crossing such as the Papago have. The Kumeyaay nation was cut in two by the Mexican-United States border just as the Papago land was cut. However, only the Papago have gained the right of free passage at the border.

September 1990
Florence C. Shipek

Professor Emeritus, University of Wisconsin Parkside
First Costo Professor of American Indian History 1987-88
University of California, Riverside.

Footnote 1: "Kumeyaay" is now the agreed-upon spelling of the tribal name for the Indians who inhabited the region extending along the coast for approximately 50 miles both north and south of the international border between Mexico and the United States and from the coast almost to the Colorado River above the border and to the river below Yuma (Quechan territory), but north of Cocopa territory. The term "Diegueño" had been attached to them by the Spanish because they were in the territory assigned by the Spanish church to Mission San Diego. The surviving members of the tribe, on thirteen reservations in San Diego County, have requested that their own tribal name be used. During my research from 1956 through 1965, the majority of those people who were over 80 years old provided "Kumeyaay" as the tribal name they called themselves, to separate themselves from Cocopa, Quechan, Paipai, Cahuilla or Kahway (their term for Luiseño). Various Spanish spelling variations of this name were also found in the Spanish records concerning the people of this region.

HER ETHNOBOTANIC CONTRIBUTIONS

For ethnobotanical identification I had taken Delfina to many locations in southern San Diego County, as I did with many Kumeyaay elders, and rewalked some areas many times, in different seasons, and after rainfalls, as was done with the other Kumeyaay. However, Delfina was the only witness that I had been able to take to Torrey Pines State Park and the Mission Bay Salt Marsh Reserve. Therefore it is appropriate to include this portion of her ethnobotanical work in this epilogue. Her plant identifications in other locations will be included in the general Kumeyaay ethnobotany under preparation and which includes the work of a number of Kumeyaay elders.

Unlike other locations, to both Torrey Pines and the Mission Bay Salt Marsh Reserve, we made only one visit. Delfina noted that in other seasons of the year, we would see additional or different useful plants. At Torrey Pines, in April, 1968, we were taken on a lengthy path through the trees by the State Park Ranger but did not go near the cliffs; from the marsh area which adjoined the park on the northeast, some plants were brought to us. Later Delfina said that there were many plants used in the past which we did not see on this trip.

At the Mission Bay Salt Marsh we walked over the firmer areas and places where we did not sink too deeply in the mud. Again, Delfina stated that additional plants

83

had been used in the past but that she had not seen them. She particularly looked for a plant with a white tuberous root which she said was like a small potato but was unable to find it. She also said that there used to be a mud worm that they would dig up and fry. We visited the reserve in June 1966 and Delfina said that usually they went to the desert in June. This could be the reason she did not find all the plants for which she looked.

It had been so long since Delfina had been in both these areas, she could not always remember the Kumeyaay name for the plants, but she could remember those her family used.

In the chart that follows, the Kumeyaay name is boldface and italics. (For pronunciation refer to the new guide by Dr. Langdon on pages 19 and 20.)

In working with the plant specialists and elders, I discovered that each had his or her own special knowledge of medicinal and food plants. Also if a plant was not used at the time we went through an area, I often was told "no use," but several months later, on another visit was told a use for the plant at that stage of its annual growth. Thus with only one visit to these locations, the information is incomplete and the words "no use" may have referred only to that growth stage, or use by Delfina's family. Botanical identifications are according to Philip A. Munz, <u>A Flora of Southern California</u> (1974, University of California Press).

Adenostoma fasciculatum (chamise, greasewood)***Hamuchi***
Torrey Pines State Park

"I never used this."

Antirrhinum nuttallianum (snapdragon) **Pullaay**
Torrey Pines State Park

> "We made tea for colds by gathering purple flowers. Boiled it and added a little oil, now olive oil, and drink."

Artemisia californica (sage) **Kuchash**
Torrey Pines State Park

> "Grind leaves and use fresh as poultice on ant bites or boil and use for tea when ill; boil and bathe in it for measles. It was dried and used as a tobacco for smoking also."

Atriplex californica (saltbush)
Mission Bay Salt Marsh Reserve

> Delfina stated that this was one of two kinds of this plant. "Grind leaves and stems to use fresh as a poultice on ant bites; also leaves are boiled for tea for stomach ache."

Atriplex semibaccata (salt weed, Australian saltbrush)
Mission Bay Salt Marsh Reserve

> Then we saw this, the second kind, which was used for food. "Kumeyaay gathered young leaves and ate them after boiling them several times to remove the bitterness. The seeds are no good."

Avena fatua (wild oats) **Nyipaay**
Mission Bay Salt Marsh Reserve

(This was grass introduced from Europe which replaced the semi-domesticated grain which was broadcast and harvested when the Spanish arrived, but destroyed by overgrazing). "Kumeyaay collected the seeds and ground them for pinole."

Batis maritima   **Millykami**
Mission Bay Salt Marsh Reserve

The leaves and stems can be chewed fresh for the water in them, or they are boiled and eaten as a vegetable.

Brassica nigra   **Hamull**
Mission Bay Salt Marsh Reserve

Plant naturalized from Europe. "Our name means greens, any greens used for food. We cooked leaves for greens. The seeds of this one were used as medicine. When ground, boiled, and strained, the liquid is used to wash eyes, for pink eyes especially."

Brodiaea sp.
Both locations; *B. capitata* at Torrey Pines

"We ate the bulbs after baking them."

Bromus carinatus   **Perhaaw**
Mission Bay Salt Marsh Reserve

"We collected the seeds and ground them for pinole."

Chenopodium sp. *Hakwach*
Torrey Pines State Park

"The roots are mashed for soap. The seeds are ground for pinole."

Chenopodium murale *Pilluull*
Mission Bay Salt Marsh Reserve

"We gathered young leaves to boil for greens; the older leaves required 2 or 3 boilings to remove bitterness. When seeds form, we gather them for pinole."

Chrysanthemum coronarium *Istap*
Mission Bay Salt Marsh Reserve

"Another slightly different form of this is in the mountains, both are used same way; boil stems and leaves and bathe in water when the body aches. We also boil seeds (without grinding) or dried flowers for tea for stomach trouble to take before or after meals. We also give the tea just before and after a baby is born."

Cneoridium dumosum (spice bush, bushrue)
Torrey Pines State Park

"Boiled, the plant was used as mouthwash and a gargle, and also for a toothache."

Convolvulus aridus (morning glory) *Mu'uch*
Torrey Pines State Park

We used it as a medicine; boiled the whole plant
and bathed sores in the liquid.

Coreopsis maritima (sea dahlia, tickseed)   ***Tesa***
Torrey Pines State Park

"Boiled roots for tea for stomach ache."

Corethrogyne filaginifolia (sand aster)   ***Kumhwaay***
Torrey Pines State Park

"Boil purple flowers then boil and drink tea for
aching chest."

Croton californicus   ***Ahwaay kahwaw***
Torrey Pines State Park, down on slough side

"Gather the leaves and flowers and whole plant,
boil it and use liquid to wash eyes, especially for
pink eye."

Cryptantha intermedia (Popcorn flower)   ***Shemap***
Torrey Pines State Park

"I do not know any use for this one."

Cuscuta salina (dodder)   ***Haakwal pehaa***
Mission Bay Salt Marsh Reserve

"The name means 'lizard's guts'. I don't know any
use."

Daucus pusillus (rattlesnake weed)
Torrey Pines State Park; down on slough side

"Boil whole plant and use as medicine for a toothache; also for fevers, drink as a tea."

Distichlis spicata (salt grass)
Mission Bay Salt Marsh Reserve and slough below Torrey Pines State Park

"Boil and use as mouth rinse when mouth is sore."

Dudleya edulis (live-forever)   *Millykumil*
Torrey Pines State Park

"We ate the fleshy parts raw. Some are salty, some sweet and some bitter. There is a bigger one that we ate also. They are best when young and fresh."

Dudleya lanceolata (live-forever)   *Millykumaay*
Torrey Pines State Park

"We ate the leaves green and raw, not cooked." Delfina had not seen this one in years and, through Mrs. Robertson, asked permission for a small cutting to transplant to her new home on the reservation. The ranger smiled and walked rapidly out of sight.

Encelia californica (sunflower)   *Nahekwi*
Torrey Pines State Park

"The name means 'it watches the sun'; I did not use it."

Eriogonum fasciculatum (buckwheat)   **Hamill**
Both locations

> "We gathered the flowers or roots and boiled them
> to drink as tea for stomach trouble; roots
> are best. The tops are eaten for food." At
> the marsh, she called the form there
> *hamillta* meaning "big *hamill*."

Eriophyllum confertiflorum (yarrow)   **Chanewan**
Torrey Pines State Park

> "This is used for someone with pimples on their
> face. They were told to boil the whole plant and
> wash face in water to clear away the pimples."

Foeniculum vulgare (sweet fennel)
Torrey Pines State Park
(naturalized from Europe)

> "We boiled the seeds to drink the liquid as tea for
> stomach aches."

Frankenia grandifolia   **Chayaaw**
Mission Bay Salt Marsh Reserve and Torrey Pines Slough

> "We used this as a medicine; made tea with the
> whole plant and drank the tea for colic."

Galium angustifolium (bedstraw)   **Hatpat**
Torrey Pines State Park

> "Used for diarrhea by boiling as a tea. It was
> gathered when green and in bloom and could be
> dried and saved till needed."

Gnaphalium bicolor (pearly live forever)   **Kumil**
Torrey Pines State Park

"Boiled and used as poultice on sores."

Haplopappus venetus   **Michashi**
Torrey Pines State Park

"Dry stalks were used for brooms."

Hedypnois cretica
Mission Bay Salt Marsh Reserve

"The whole seeds (not ground) are boiled for tea to drink for stomach trouble."

Helianthemum scoparium (rock rose)
Torrey Pines State Park

"Boil yellow flowers for tea to give to mothers having a difficult birth."

Heliotropium curassavicum   **Millykupish**
Mission Bay Salt Marsh Reserve

"The roots are boiled for medicinal tea to regulate menstruation."

Hemizonia ramosissima (tarweed)   **Hatuun**
Mission Bay Salt Marsh Reserve

"We boiled whole plant for steam when someone has a headache. Formerly we used it in sweathouses; now put it in a pan and put a towel over your head."

Heteromeles arbutifolia (toyon)   **Huuchih**
Torrey Pines State Park

> "Make a pulp of the leaves and wash sores with the
> liquid. The berries were bitter and used for food
> only when we were starving."

Isomeris arborea (bladderpod)   **'epshash**
Torrey Pines State Park

> "We eat flowers after they are cooked and drained
> several times to remove the bitterness."

Jaumea carnosa
Mission Bay Salt Marsh Reserve

> "There are two kinds, one with smell is best; this
> one has no smell and is not as good and not used.
> The other kind used to be in the marsh but I can't
> find it. It was boiled as a tea for fever; it was also
> cooked and eaten as a vegetable."

Limonium californicum
Mission Bay Salt Marsh Reserve

> "We boiled the young leaves to eat as vegetables.
> The leaves also can be dried and stored, they keep
> well for future use."

Limonium sinuatum (sea-lavender)
Mission Bay Salt Marsh Reserve

> "We boil leaves for a tea to take for diarrhea.
> Generally we saw the one with blue flowers
> instead of white like this.

Malva parviflora (cheeseweed) **Mal**
Mission Bay Salt Marsh Reserve

> "We made medicinal tea from the dried buds for drinking when feverish; also boiled the whole plant for a bath. As food, we boiled the young leaves for vegetables."

Marah macrocarpus (wild cucumber, big-root, chilicothe)
Torrey Pines State Park

> "We ground the black seeds, mixed them with water and used the black as makeup. As medicine, we boiled the leaves to use on hemorrhoids."

Mesembryanthemum chilense (sea fig) **Hayaaw**
At both locations

> "We ate fruit fresh; leaves were as a vegetable; seed was ground for pinole."

Mesembryanthemum crystallinum (ice plant) **Sii'ii nesii**
Both locations

> "Used red berries and red leaves as face rouge and paint; roots or whole plant was ground for soap. Away from the salt marsh, leaves are cooked and eaten as greens; at the marsh or beach they are too salty to eat."

Mesembryanthemum edule (hottentot fig) **Hayaaw**
Mission Bay Salt Marsh Reserve

> "We boiled and ate young green parts, just like young green shoots of cactus were boiled."

Mimulus puniceus (monkeyflower)
Torrey Pines State Park

> "We boiled the plant for tea to regulate menstrual periods. It can be used fresh or dried and stored."

Mirabilis californicas (four o'clock)   **Meshkatull**
Torrey Pines State Park

> "We used the root, flower and the whole plant to make a tea to drink for stomach aches."

Opuntia occidentalis (prickly pear)   **Melltat**
Torrey Pines State Park

> "We eat the fresh fruit and fry or boil the young green pads. All types of Opuntia were eaten in these ways."

Pinus Torreyana (Torrey Pine)   **'ehwiiw**
Torrey Pines State Park

> "The name means 'pine nut.' The pine nuts are generally collected in September (when ripened), sometimes the cones had to be roasted to get the seeds out. Eaten as nuts raw or roasted; they are also ground and cooked as pinole or added to other seed flours for flavoring."

Polypodium californicum (fern, Polypody)   **'awi hatat**
Torrey Pines State Park

> "Boil the roots and leaves to use for internal bleeding. The name means "rattlesnake's back;" that's what it looks like."

Raphanus sp. (radish)   *Hamull*
Mission Bay Salt Marsh Reserve

> "We cooked the young leaves and stems for greens.
> For medicine, we grind the seeds, boil and strain
> to use the liquid to bathe inflamed eyes, and for
> pink eyes."

Rhamnus crocea (redberry)   *Tat*
Torrey Pines State Park

> "When someone had a captive mocking bird,
> gathered these to feed to the bird. We kept the
> birds for their songs."

Rhus integrifolia (lemonadeberry)   *Huusill, Huutat*
Torrey Pines State Park

> "The name used depended upon the berry color red
> (*huusill*) or orange (*huutat*); the berries were eaten
> fresh or soaked to flavor water; the seed (*keha*)
> was ground and used with fruit for tea. We also
> ground the seeds to drink when sick and feverish.
> The bark was also made into a tea to use after a
> baby was born."

Rhus laurina (Laurel Sumac)   *'ektii*
Torrey Pines State Park

> "The bark was used for tea after the birth of a
> baby. Some used it for venereal diseases also."

Rosaceae sp. (wild rose)   *kwa'ak*
Torrey Pines State Park down on slough side

"The wild rose was used as a medicine; gather flower petals or leaves (if no flowers available); boil them and use liquid to bathe eyes when have a cold in the eyes. The fruit and seed ground as food."

Rumex crispus (curly dock)   **Kish**
Torrey Pines State Park

"We ate the young leaves boiled as greens. When the plant is old, gathered only the seed to grind on a metate for pinole."

Salicornia pacifica and
Salicornia virginica (glasswort)   **Semull**
Mission Bay Salt Marsh Reserve

"Some people chew them for the salt."

Salvia apiana (white sage)   **Pestaay**
Torrey Pines State Park

"We ground the seeds for pinole; gathered young branches (before the flowers come) to dry, crumble and store to make tea for chest colds and coughs. Heat and smell steam when congested."

Salvia mellifera (black sage)   **Ha'anya yul**
Torrey Pines State Park

"We used it for a medicine. The leaves and stems could be used fresh or dried. They were boiled and the water used for bathing when a person ached due to flu, rheumatism and arthritis."

Sanicula arguta (cow parsley, sanicle, snakeroot)　　*Chap*
Torrey Pines State Park

> "Roots are boiled to eat; leaves are boiled as tea for cramps."

Sisyrinchium bellum (blue eyed grass)　　*Michkal*
Torrey Pines State Park

> "Boil whole plant as tea for cramps."

Solanum xantii (purple nightshade)　　*'ewii eyiiw*
Torrey Pines State Park

> "We gathered this, dried and ground the whole plant to put on feet for athletes foot. The name means 'snake eyes'."

Spartina foliosa (marsh grass)　　*Tapish*
Mission Bay Salt Marsh Reserve

> "When this grows big, we made them into bundles for house walls. Other uses were as medicine; boil roots and give to baby when constipated or unable to urinate."

Stephanomeria virgata (wire lettuce)　　*Telkuu*
At both locations

> "Gather the whole plant, it can be dried and saved. Boil the roots and drink liquid to get rid of intestinal worms. From Mexicans, we learned to boil whole plant to clean stomach after a hangover."

Suaeda Torreyana (sea-blite, seep-weed)
Mission Bay Salt Marsh Reserve

>"We crushed the whole plant to use for soap."

Urtica holosericea (nettle)   **Hampasis**
Torrey Pines State Park, down in the slough

>"Formerly we gathered and boiled for greens. We also cooked a lot to a real strong liquid and bathed in it when we got into poison oak, and also for skin diseases."

Xylococcus bicolor (manzanita)   **Haasill**
Torrey Pines State Park

>"Soak ripe berries and use as a cool drink. Put in water jar so that evaporation cools the drink."

Yucca schidigera (Spanish bayonet, Mohave yucca) **Sha'a**
Torrey Pines State Park

>"The roots are mashed and used for soap. The leaves were used for fibers in many ways. Leaves were split into narrow strips and used to tie houses together; or strips braided to make pottery rests. Leaves were shredded to fibers to make sandals, or quick containers which were used and then thrown away. Flower petals were eaten raw when young and tender; they were boiled twice and the water thrown off when the petals were older. Some people did not eat the flowers, others did. We strung seeds for beads; chopped them for tea, or ground to cook as mush."

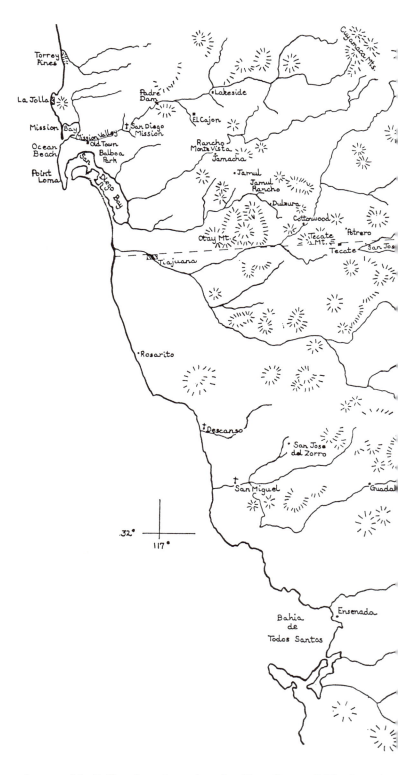

Territory known and traversed by Delfina Cuero in southern San Diego County, California, and northern Baja California. Drawn by Florence C. Shipek. Although the majority of streams in this

CALIFORNIA
BAJA CALIFORNIA

Mexicali

La Rumerosa

Laguna Maquata

Sierra de Cocopa

Tres Pozos

Mesquite

Sierra Juarez

San Pedro

San Juan

La Huerta

‡ Santa Catarina

116°

region are intermittent, they have been represented by solid lines in order to indicate the overall geographic unity of the areas immediately north and south of the International Boundary.

OTHER BALLENA TITLES
(Prices subject to change)

Bean, Lowell John, ed. SEASONS OF THE KACHINA: PROCEEDINGS OF THE CALIFORNIA STATE UNIVERSITY, HAYWARD CONFERENCES ON THE WESTERN PUEBLOS 1987-1988. 1989. BP-AP, No. 34. ISBN 0-87919-115-5, clothbound, $32.95; ISBN 0-87919-114-7, paperbound, $21.95.

Bean, Lowell John, and Thomas C. Blackburn, eds. NATIVE CALIFORNIANS: A THEORETICAL RETROSPECTIVE. 1976. ISBN 0-87919-055-8, paperbound, $17.95.

Bean, Lowell John, Sylvia Brakke Vane, and Jackson Young. THE CAHUILLA INDIANS: THE SANTA ROSA AND SAN JACINTO MOUNTAINS. 1991. BP-AP No. 37. ISBN 0-87919-120-1, paperbound, $14.95; ISBN 0-87919-121-X, clothbound, $19.95.

Blackburn, Thomas C., and Travis Hudson. TIME'S FLOTSAM: OVERSEAS COLLECTIONS OF CALIFORNIA INDIAN MATERIAL CULTURE. 1990. BP-AP No. 35. ISBN-0-87919-116-3, paperbound, $24.95; ISBN 0-87919-117-1, clothbound, $34.95.

Chamberlain, Von del. WHEN STARS CAME DOWN TO EARTH; COSMOLOGY OF THE SKIDEE PAWNEE INDIANS OF NORTH AMERICA. 1982. BP-AP No. 26. ISBN 0-87919-098-1, paperbound, $17.95.

Ericson, Jonathan E., R. E. Taylor, and Rainer Berger, eds. PEOPLING OF THE NEW WORLD. 1982. BP-AP No. 23. ISBN 0-87919-094-9, paperbound, $19.95.

Great Basin Foundation. WOMAN, POET, SCIENTIST: ESSAYS IN NEW WORLD ANTHROPOLOGY HONORING DR. EMMA LOUISE DAVIS. BP-AP NO. 29. ISBN 0-87919-106-6, paperbound, $30.00.

Heizer, Robert F. FEDERAL CONCERN ABOUT CONDITIONS OF CALIFORNIA INDIANS, 1853-1913. 1979. ISBN 0-87919-084-1, paperbound, $9.95.

Hudson, Travis, and Thomas C. Blackburn. MATERIAL CULTURE OF THE CHUMASH INTERACTION SPHERE, 5 volumes, 1982-1987. BP-AP Nos. 25, 27, 28, 29, and 31. ISBN 0-87919-100-7, paperbound; ISBN

0-87919-101-5, clothbound, $190.00. Prices of individual volumes on request.

Hudson, Travis, and Ernest Underhay. CRYSTALS IN THE SKY: AN INTELLECTUAL ODYSSAY INVOLVING CHUMASH ASTRONOMY, COSMOLOGY AND ROCK ART, 1978. BP-AP No. 10. ISBN 0-87919-074-4, paperbound, $18.95.

Jewell, Donald P. INDIANS OF THE FEATHER RIVER: TALES AND LEGENDS OF THE CONCOW MAIDU OF CALIFORNIA. 1987. ISBN 0-87919-111-2, paperbound, $12.95.

Meighan, Clement W., and V. L. Pontoni, eds. SEVEN ROCK ART SITES IN BAJA CALIFORNIA. 1979. ISBN 0-87919-081-7, paperbound, $10.95.

Miller, Virginia P. UKOMNO'M: THE YUKI INDIANS OF NORTHERN CALIFORNIA. BP-AP No. 14. ISBN 0-87919-083-3, paperbound, $8.95.

Stewart, Irene. A VOICE IN HER TRIBE: A NAVAJO WOMAN'S OWN STORY. 1980. BP-AP No. 17. ISBN 0-87919-088-4, paperbound, $8.95.

Stickel, Gary E., ed. NEW USES OF SYSTEMS THEORY IN ARCHAEOLOGY. 1982. BP-AP No. 24. ISBN 0-87919-096-5, paperbound, $9.95.

Sutton, Mark Q. INSECTS AS FOOD: ABORIGINAL ENTOMOPHAGY IN THE GREAT BASIN. 1988. BP-AP No. 33. ISBN 0-87919-139-2, paperbound, $17.95.

Vane, Sylvia Brakke, and Lowell John Bean. CALIFORNIA INDIANS: PRIMARY RESOURCES, A GUIDE TO MANUSCRIPTS, ARTIFACTS, DOCUMENTS, SERIALS, MUSIC AND ILLUSTRATIONS. 1990. BP-AP No. 36. ISBN 0-87919-117-1, paperbound, $33.00; ISBN 0-87919-117-1, clothbound, $45.00.

Wilkie, Philip J., ed. BACKGROUND TO PREHISTORY OF THE YUHA DESERT REGION. BP-AP No. 5. ISBN 0-87919-058-2, $7.95.

Orders to:
Ballena Press Publishers' Services, P.O. Box 2510, Novato, CA 94948.
Telephone 415-883-3530.